Technical Drawing Presentation & Practice

Segun R. Bello
[MNSE, R. Engr. COREN]

Createspace.com

Technical Drawing
Presentation & Practice

Copyright © 2012 by Segun R. Bello

Federal College of Agriculture Ishiagu, 480001 Nigeria
segemi2002@gmail.com; bellraph95@yahoo.com
http://www.dominionpublishingstores.yolasite.com
http://www.segzybrap.web.com
+234 8068576763, +234 8062432694

No part of this book may be reproduced, stored in a retrieval system or transmitted, in any form or by any means, electronics, mechanical, photocopying, recording or otherwise, without the prior written permission of the copyright holder.

DPs Dominion
Publishing Services

ISBN-13: 978-1481250122
      10: 1481250124

First published in December 2012
Printed by Createspace US

Createspace
7290 Investment Drive
Suite B North Charleston,
SC 29418 USA
www.createspace.com

Dedication

To the glory of God Almighty

Acknowledgements

Glory be to God Almighty, the author of life and the giver of knowledge. I acknowledged the several authors and researchers whose wealth of experiences documented and made available in book and journal prints forms, as well as numerous materials available of the web.

Many thanks to all students, past and present, that had passed through my tutelage as instructor and teacher for their contributions, criticisms and feed-backs on the series of classroom lectures and practical experiences which are documented in this book.

Special thanks go to my wife, who had always back-up the realization of God's plan for me. She is a virtuous woman in class of her own. I am grateful for her understanding and tolerance in taking full responsibility of running our home during the scripting and review of this works.

I am grateful to my children, Ayomikun, Pelumi, Damilola and Adeola, who have been so wonderful and cooperative during this period. I am encouraged and strengthened by their prayers, my God shall surely reward them. Amen

Content

Dedication iv
Acknowledgements v
Preface x
SECTION 1   Drawing Presentation xii
Chapter 1   Introduction to Technical Drawing 14
1.1   Introduction 14
1.2   Introduction 14
1.3   Drawing language and standards 15
1.4   Aim of drawing 16
1.5   Types of drawing and views 16
1.6   Engineering drawing requirements 17
1.7   Professional application of technical drawing 18
1.8   Technical drawing and agricultural development 18
1.9   Steps to planning drawing 19
Exercise 19
Chapter 2   Drafting and Drawing Presentation 20
2.1   Introduction to drafting 20
2.2   Drafting tools 20
2.3   Importance of drafting 21
2.4   Drafting practices 22
2.5   Drafting aids/instruments 23
2.6   Using the instruments 35
2.7   Drawing reproduction 44
2.8   Scales in drawing 45
2.9   Components of drawing 52
Exercise 53
Chapter 3   Lettering Principles and Practice 57
3.1   Introduction 57
3.2   Importance of lettering 57
3.3   Types of lettering 57
3.4   Style of lettering 58
3.5   Lettering fonts in drawing 58
3.6   Lines in lettering 58
3.7   Lettering practice 59

3.8	Drawing lettering lines	60
Exercise		61
Chapter 4	Principles of Dimensioning	63
4.1	Importance of dimensioning	63
4.2	Guidelines for dimensioning	63
4.3	Types of dimensions	63
4.4	Components of dimension	64
4.5	Dimension placement	64
4.6	Rules of dimension	67
4.7	Types of dimensioning	69
4.8	Dimensioning best practices	78
Exercise		79
SECTION 2	**Drawing Geometry & Projections**	82
Introduction		84
Chapter 5	Geometrical Figures: Properties and Construction	86
5.1	Introduction	86
5.2	Point	86
5.3	Lines, properties, application and construction	87
5.3.1	Line styles	88
5.3.2	Forms of lines	89
5.3.3	Types of lines	89
5.3.4	Line applications	91
5.3.5	Line drawing	92
5.3.6	Line construction	93
5.3.7	Loci: Properties and construction	100
5.4	Plane and solid loci	103
5.5	Plane figures	104
5.5.1	Angles and their properties	105
5.5.2	Triangles, their properties and construction	107
5.6	Quadrilaterals, their properties and construction	113
5.7	Polygon, their properties and construction	116
5.8	Circles, their properties and construction	123
5.8.1	Circle geometry	126
5.8.2	Tangency and normalcy construction	133
Exercise		137
Chapter 6	Projections in Engineering Drawings	142
6.1	Drawing presentation	142

6.2	Features of projection	144
6.3	Projection of points	146
6.4	Projection of lines	146
Chapter 7	Descriptive Geometry	153
7.1	Introduction	153
7.2	Orthographic projection	153
7.3	Multiview projection	155
7.4	Principles of first angle orthographic projection	163
7.5	Principles of third angle orthographic projection	168
7.6	Symbols for orthographic projection	171
7.7	Comparing 1st and 3rd angle projection	172
6.5	Auxiliary projection	172
Exercise		180
Chapter 8	Pictorial Drawing and Construction	182
8.1	Introduction	182
8.2	Projection of pictorial drawing	182
8.3	Isometric drawing	183
8.4	Oblique drawing	188
8.5	Axonometric drawing	190
8.6	Perspective drawing	191
8.7	Diametric drawing	196
8.8	Model building drawings	196
8.9	Assembly drawings	198
8.10	Sketch drawings	200
8.11	Comparing projections	207
Exercise		208

SECTION 3   Conic Sections & Surface Development 211

Introduction		213
Chapter 9	Conic Sections and Construction	214
9.1	Introduction	214
9.2	Construction of conic sections	217
9.2.1	Ellipse and its construction	217
9.2.2	Hyperbola and its construction	231
9.2.3	Parabola and its construction	235
Exercise		241
Chapter 10	Solid Sections and Development	243
10.1	Introduction	243

10.2	Types of sections	246
10.3	Hatching	250
10.4	Sections of solid	253
10.5	Surface development	261
10.5.1	Full surface development of geometrical solids	262
10.5.2	Development of lower surfaces of geometrical solids	264
10.5.3	Interpenetration of surfaces	270
Exercise		276
References		278
Notes		279

Preface

Engineering/technical drawing is a pre-requisite course for all who wish to pursue a career in engineering profession especially in design and construction program. Emphasis is placed on media drafting, lettering, and alphabet of lines, geometric construction, sketching, and multiview drawings. Students learn traditional drafting techniques through the study of geometric construction at which time they are introduced to computer aided drafting and design.

This book is therefore designed to help students acquire requisite knowledge and practical skills in engineering/technical drawing practices. The contents were designed to prepare students for technical, diploma and degree examinations in engineering, engineering technology and technical vocations in other professions in the monotechnics, polytechnics and universities.

Therefore, when armed with this book, students should be able to;

- Understand the principles and techniques of drawing, presentation and projections
- Understand drawings and its applications to geometry; plane or solid
- Understand the principles and application of free hand sketching
- Understand development of surfaces and conic-sections

At the end of each chapter are lists of practical exercises that will help students perfect their skill and proficiency in technical works.

Segun R. Bello
480001, Nigeria

SECTION 1

Drawing Presentation

Chapter 1

Introduction to Technical Drawing

1.1 Introduction

One of the best ways to communicate one's ideas is through pictures, graphic illustration or drawings. Details of engineering innovations and technical inventions are hid in drawing for the purpose of safeguarding them. Technical/engineering drawing is a means of communicating shapes, sizes, positions and proportion, features and precision of physical objects.

The following descriptions can be used to describe the field of technical/engineering drawing and presentation:

- Engineering drawing is graphical representation of physical objects and their relationship.
- It is a universal language of engineering used in design processes for solving problems quickly and accurately by visualizing objects and conducting analysis.
- It can also be said to be a graphic representation of objects and structures used to solve problems which involve special relationships.
- It is a mode of thinking in which two-dimensional projections are used to visualize three-dimensional situations.
- It is also a means of describing and defining processes which verbal expressions cannot adequately conveyed.
- It is therefore an extension of language, and as such, an essential part of education in a technological society.
- Engineering drawing communicate product design and manufacturing information in a reliable and unambiguous manner regardless of language

1.2 Introduction

Drawings are used in all fields of engineering (agriculture, mechanical, civil, architectural, electrical, aerospace, etc.). The types of drawings we will be creating in this book are mechanical drawings, but the concepts are all transferable to the other engineering fields. Engineering drawing is concerned with imparting precise information hence it is understandable that neatness and accuracy should play an important part in its practice. Beautifully copied engineering drawings are of little value if the principles behind the work are not fully understood and applied.

These principles can be learned and practiced by using freehand, mechanical, or computer aided design (CAD) methods. The ability to read drawing is the most important requirement of all technical people in engineering profession. Below are the basics concepts of engineering drawings.

1.3   Drawing language and standards

Engineering drawing is a form of language in its own right with rules and signs. Just as it is applicable to any language, certain rules (or standards) must be followed in producing any drawing. These rules/standards define how shapes and position of object should be represented; for instance, the order of orthographic views and different line types has rules and specific position in order to fully describe such object. They also define how a part should be dimensioned or tolerance.

These standards are developed by some institutions or governing agencies specifically set up for the purpose of formulating standards globally accepted and subject to update on a 5 year basis. These organizations or governing agencies include:

1. BSI - British Standards Institute
2. ASME-American Society of Automotive Engineers
3. ANSI- American National Standards Institute
4. DIN - Deutsches Institut fur Normung (Germany)
5. ISO - International Standards Organization

The governing agency responsible for setting the mechanical drawing standards and practices used in creating technical drawings of mechanical parts and assemblies is the American Society of Automotive Engineers (ASME).Considering the ASME standards for example, there are a number of documents published by ASME that cover various aspects of mechanical drawings, here are a few of them:

1. ASME Y14.100 -2004 *Engineering drawing practices*
2. ASME Y14.1 -1995 *Decimal inch drawing sheet size and format*
3. ASME Y14.3M *–Multi and sectional view drawings*
4. ASME Y14.4M -1989 *Pictorial drawing*
5. ASME Y14.5M –1994 *Geometric dimensioning and tolerance*
6. ASME Y14.13M -1981 *Mechanical spring representation*

It is important to follow these standards to ensure your drawings are interpreted correctly by others. Always consult the standard when in doubt!

1.4 Aim of drawing

The main purpose of engineering drawings is to communicate to other engineers, machinists, etc. Drawings do the communication best merely because a picture is worth a thousand words. Giving all of the information needed to make the product and being accurate in that information is the main goal. Engineers are very picky about their drawings and must pay attention to detail.

1.5 Types of drawing and views

There are terms commonly associated with graphic and engineering design drawings in various forms and are meant to express different ideas as indicated below:

Diagram: This type of drawing depicts the function of a system represented in drawing form

Sketching: This generally refers to freehand drawing without the aid of drawing instrument

Drawing: This term usually means using instruments or drawing aids ranging from compasses to computers to bring precision to an expressed conception in form of graphics.

Drawing list: This is the list of cross references drawings that all combined to produce an single product

Parts list (bill of materials): Part listing in drawing shows material, number/quantity and provides reference number of various components

Assembly drawing: This shows how an individual parts are combined, refers to parts list

Design layout drawing: This represents broad principles of feasible solution

Arrangement drawing: This type of drawing shows finished arrangement of assemblies, including functional and performance requirements

Detail drawing: This is a single part drawing containing all information for fabrication. When there is a great disparity between feature sizes, or views are overcrowded with dimensions, a detail view can be used to capture the feature(s) of interest and display them in a removed view of greater scale.

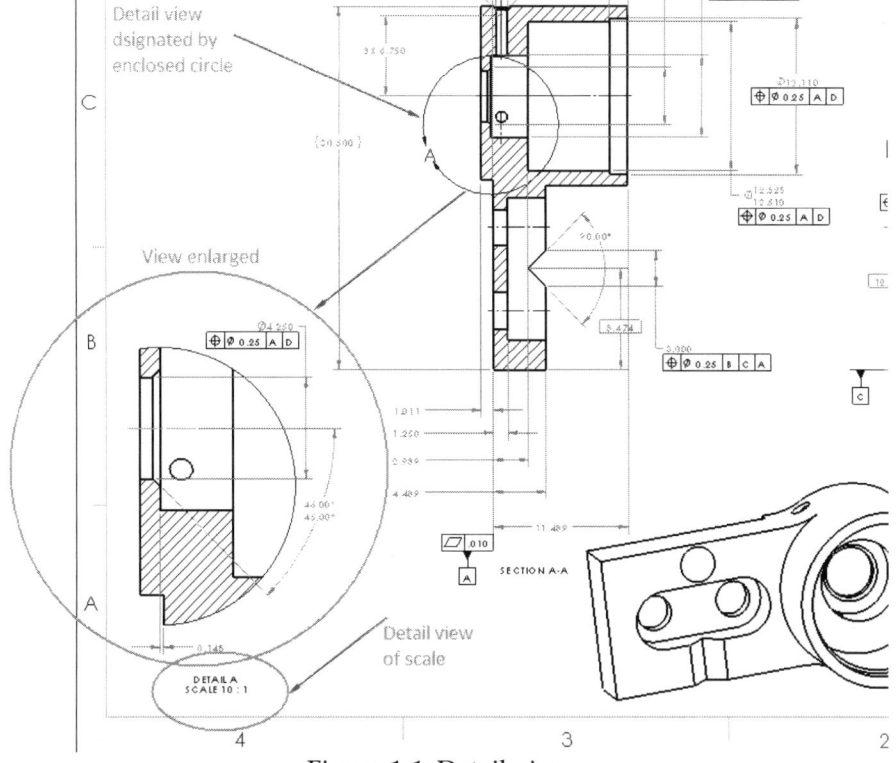

Figure 1-1: Detail view

1.6 Engineering drawing requirements

Engineering drawing requirement are conditions that a specific drawing must meet in order to conform to global standards in such field. Such requirements include:

1. *Unambiguity and clarity*: All engineering drawings must be unambiguous and clear. Only one interpretation is possible.
2. *Completeness*: Every drawing must provide all information for all stages of manufacture. i.e., detailed drawings, assembly drawings, bill of materials
3. *Suitability for duplication*: Such drawing must be suitable for duplication. It must have suitable scale and clarity such that the drawing can be copied – even micro copied – without losing quality.

4. *Language independent*: The drawing must be language or Words independent and should only be used in the title block; words should be replaced by symbols.
5. *Conformity to standards*: Your drawing must be conformable to known standards. Highest standards are ISO as numerous countries learn these rules.

1.7 Professional application of technical drawing

Figure 1-2 below give a list (not limited to the displayed groups) of professional groups that are directly linked to engineering graphics in the execution of their daily routing assignment.

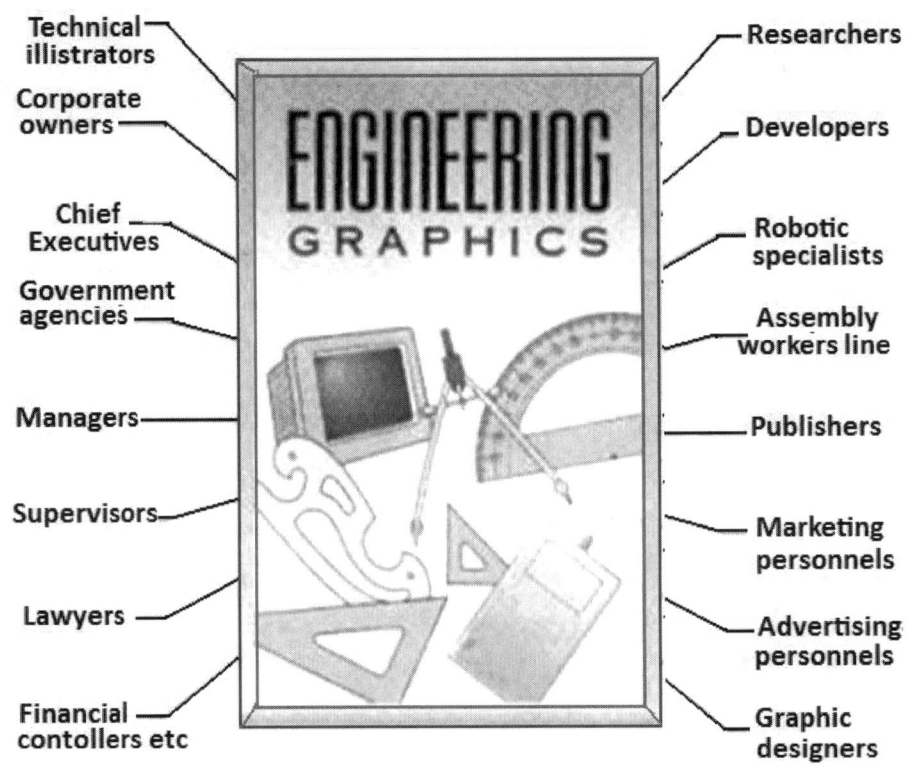

Figure 1-2: Engineering drawing partners

1.8 Technical drawing and agricultural development

In the following areas, technical drawing plays very vital roles in agricultural production practice.

1. *Farm planning and field layout*: Technical drawing provides the knowledge of mapping out and field measurements when laying out foundation of structures.
2. *Surveying:* The knowledge of the topography of an area help in the determination of the type of farming system to adopt.
3. *Employment opportunities* are created for draughtsman in the farm due to their knowledge of technical drawing.
4. *Farm structures design* and construction were gained through the basic knowledge of technical drawing
5. *Implement design*: Preliminary sketch of farm implements and tools aids their design and construction in the workshop.
6. *Machinery repairs:* Faults diagnosis and repairs on some machinery can only be done by tracing the diagram of the design details on the machine layout.

1.9 Steps to planning drawing

Unplanned drawing makes interpretation cumbersome and ambiguous bearing in mind that drawing materials could be costly. Therefore starting engineering drawing you should plan how to make the best use of your space. It is important to think about the number of views your drawing will have and how much space you will use of the paper. Consider the followings:

- Try to make maximum use of the available space.
- If a view has lots of detail, try and make that view as large as possible. If necessary, draw that view on a separate sheet.
- If you intend to add dimensions to the drawing, remember to leave enough space around the drawing for them to be added later.
- If you are working with inks on film, plan the order in which you are drawing the lines. For example you don't want to have to place your ruler on wet ink

Exercise

1. The practice of draughtsmanship is critical to the development of the agricultural sector. Critically justify this statement
2. In planning your routine daily schedule the place of graphic design and illustration cannot the overemphasis. Enumerate some of the ways this will enhance your plan.
3. Make a list of some engineering drawing standards applicable to your profession and how each is applied to effective professional practice.

Chapter 2

Drafting and Drawing Presentation

2.1 Introduction to drafting

Drafting is the art of drawing a design for something useful, ranging from tools to furniture to whole buildings. Drafting tools have developed from simple hand-held instruments like rulers and protractors into sophisticated computer-aided design (CAD) programs. Throughout history, drafting tools has evolved along with discoveries of mathematical principles.

Drafting tools are an architect/engineer's best friend. You can't begin to apply your advanced drafting skills until you've learned how to select drafting pencils and what to use, which kind of pencil for what type of line application, how to set up your paper on your drafting board and how to use dividers among several other processes and tools.

2.2 Drafting tools

Innovations in mathematics have helped drafting tools evolve into drafting devices for accurately representing an item or structure. To prepare a drawing, the use of manual drafting instruments or computer-aided drafting or design (CAD) is required. Examples of such devices include the straightedge, compass, dividers, T-square, protractor among several others. The basic drawing standards and conventions are the same regardless of what design tool you use to make the drawings.

In learning drafting, we will approach discussions on drafting tools from the perspective of manual drafting. If the drawing is made without either instruments or CAD, it is called a freehand sketch.

Different types of pens have been developed over the years to enable designers to draw different-size lines or consistently make the same lines on a drawing. Blueprints developed out of an invention by Sir John Herschel in 1842, allowing the duplication of the same drawing was popular in drawing representation until the late 20th century, when manual drafting instrument took the center stage in drafting and lately, the computer-aided drafting (CAD) programs which have become commonplace.

Early drafting tools

The first known drafting tools were the rocks and other implements used to decorate the walls of caves. The development of rice paper by the Chinese allowed the transportation of drawings from one location to another. A drafting board dating from the Babylon of 3,000 B.C.E. has been discovered. Animal skins were used for drawings and designs until the 19th century.

Drafting furniture: Drafting furniture is specifically designed to give drafters the biggest advantage in their drawing. Tables are angled to allow drafters to get the best view of their work and to be able to draw more comfortably without having to lean forward too far or sit up too straight. A sturdy chair that can be adjusted to meet the drafter's needs is also important to have. Many drafters also prefer to have a lamp that attaches to the table for additional lighting.

Drafting paper: To create a blueprint **plan**, **a** drafter who work**s** by hand **uses a** special type of paper known as drafting paper. Three different types of paper are used in drafting.

a. *Bond paper* is very similar to paper that is used in the typical office. It is the least expensive paper and comes in varying weights.
b. *Mylar paper* is plastic in nature and allows for easy erasing, even with ink. This type of paper is more durable than bond paper and is somewhat transparent.
c. *Vellum paper* is also more durable than bond paper and allows for repeated erasing of pencil lines without causing damage. This type of paper is less expensive than Mylar paper.

2.3 Importance of drafting

Drawings are essential for

1. Expressing an idea conceived through graphic illustrations
2. Completing engineering designs,
3. Planning buildings, construction work etc,
4. Estimating the quantities of materials and relative costs of project undertaking, and
5. Finally to communicate to the artisan all of the information that the designer has developed.

2.4 Drafting practices

Drafting practices had advanced to a level where standards must be conformed to. Before starting to draw, rule of thumb expects that one should estimate how large the drawing will be and centered appropriately on the page (drawing layout area). A worthwhile aid to include is a small figure identifying the location of a detail drawing in relation to the master plan.

If text is to be written on the drawing, it will normally be placed on the right or the bottom part of the drawing. The text is used to explain symbols, methods of notation and abbreviations used in the drawing. It is also possible to give directions about materials, designs, surface treatments, assembly locations, etc.

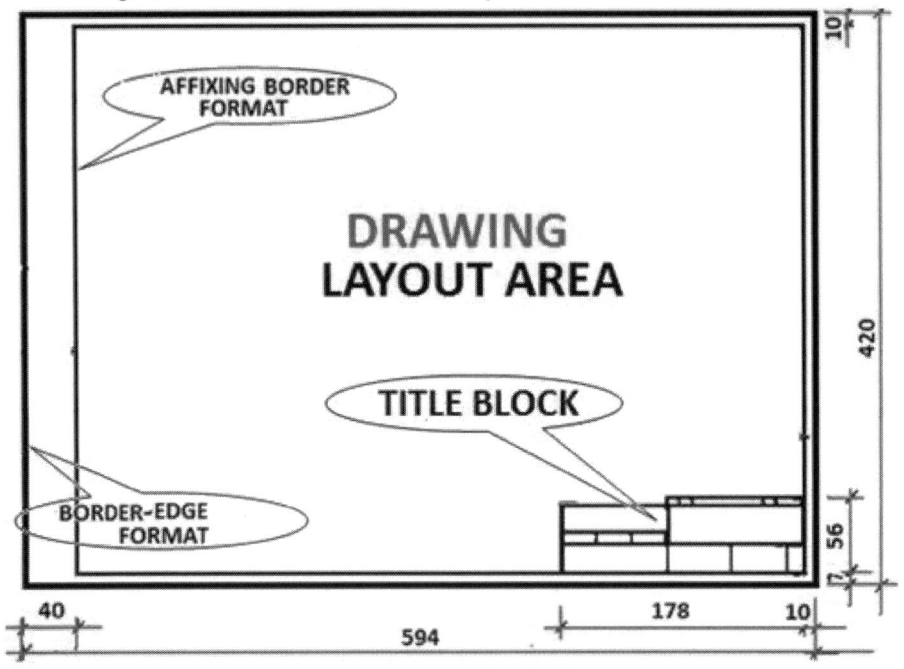

Figure 2-1: Drawing border formats

Drawings should always have borders and title boxes as shown in Figure 2-1. The wide border on one side allows several drawings to be bound together. The title box provides identification of the drawing, the designer, the draftsman and a date. The revision table above the box keeps an accurate record of all revisions.

The title box should be visible on the folded print and it should be possible to unfold the print without taking it out of the binder. The original drawing should never be folded!

2.5 Drafting aids/instruments

Technical drawing instruments are the tools used by professional and students to render precision graphics needed to manufacture a particular product or structure. These instruments take many forms because of the variety of lines and graphics needed for designs. Some instruments are manual, while others are computer-based. All professional-quality drafting instruments are manufactured with precision because the drawings they're used to make must be precise. Two types of drafting aids/instrument are in common use; the manual drafting tools and the drafting machine (drafter).

1. *Manual drafting tools for technical drawing (drawing sets)*

These are the instruments used in producing the required drawings and figures. Such instruments range in types and manufacture and the feature of each is discussed below:

a. *Drawing sheet*: Because engineering drawings include many details, they should be large enough to be accurately executed and easily read. The standard formats from the A-series should be used for all engineering drawings. However, several detail drawings may be put on one sheet. Cartridge paper is generally used in earlier stages of construction, but geometrical drawings are sold in A-size sheets (Figure 2-2).

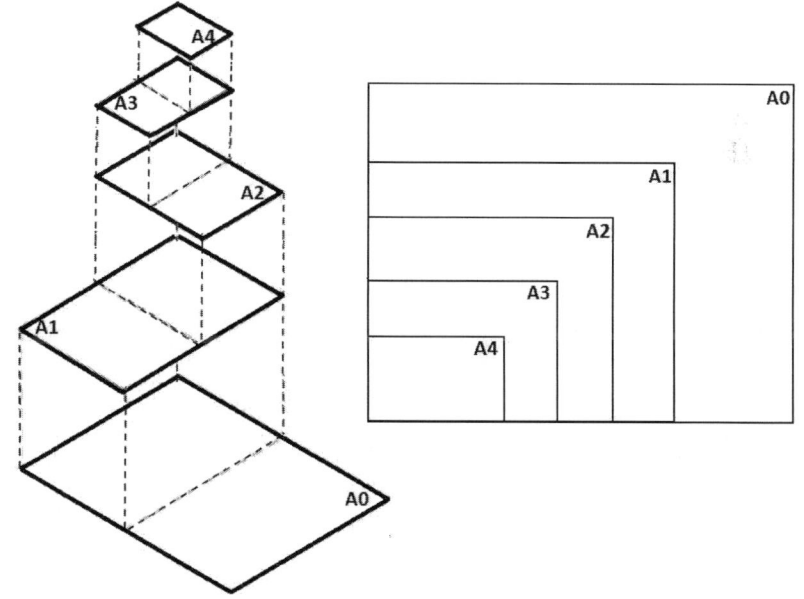

Figure 2-2: A-size paper relationships

There exist standardized sheet formats for creating engineering drawings. The following are the International Standard Organization (ISO) 'A' sheet sizes on a comparative scale:

    A0 -   1189mm x841mm

    A1 -   841mm x 594mm (the standard size of student drawing sheet)

    A2 -   594mm x 420mm

    A3 -   420mm x 297mm

    A4 -   297mm x 210mm

If the drawing plan tend to be very long, one of the following alternative paper sizes may be useful:

    A10   594 x 1189mm

    A20   420 x 1189mm

    A21   420 x 841 mm

    A31   297 x 841mm

    A32   297 x 594mm

American National Standard recommends the following drawing sheet format

        A   8.5"x 11"

        B   11"x 17"

        C   17"x 22"

        D   22"x 34"

        E   34"x 44"

If possible, only one format should be used for all drawings in a project or alternatively all drawings should have the same height. The formats A0, A10 and A20 are difficult to handle because of size of paper and should therefore be avoided. One should instead try to use a smaller size/scale or divide the figure into more drawings.

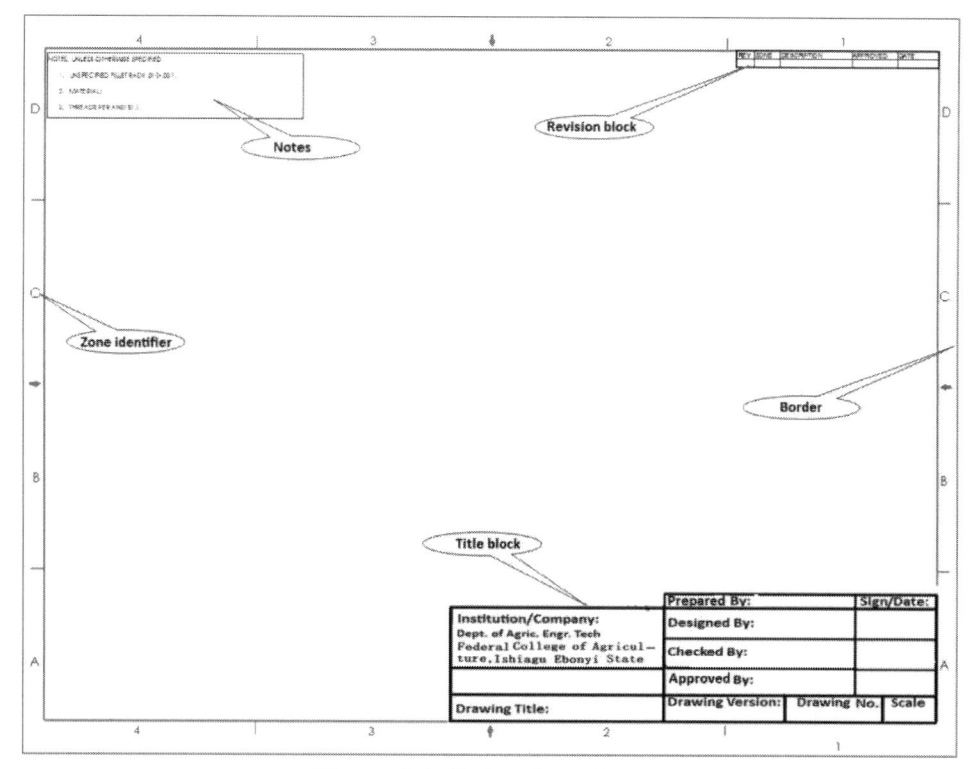

Figure 2-3: C-size sheet format

b. *Drawing board*: Obviously a good drawing board, large enough to hold the size of paper selected, is essential. One of the following sizes should be suitable:

        A0   920 x 1270mm

        A1   650 x 920mm

Figure 2-4: Drawing board

While a sheet of hardboard or blackboard may be used as a drawing board, it is advisable to install a hardwood edge such as ebony. It may be necessary to saw longitudinal grooves 75 to 100mm apart at the back of the board to prevent warping. T-squares are used to draw straight lines. The head of the square, the cross member, is placed along the left edge or the top of the drafting table, while the square's blade is laid across the table's top, over the drawing paper.

The board may be placed on a table or on trestles. The board should be covered with thick white paper or special plastic to make a smooth surface.

Figure 2-5: Drafting table

c. *Backing sheet*: Rough, pitted, scored or damaged drawing board surfaces could produce poor quality work and as such not acceptable. Such surfaces must therefore be covered with backing sheet e.g. Cardboard paper (for smooth and neat pencil work). The actual drawing sheet should be placed on top of the backing sheet.

d. *T-Square:* This is a piece of long ruler with a stock head and tapered trunk with an ebony or plastic edge. T-squares are used to draw straight horizontal lines.

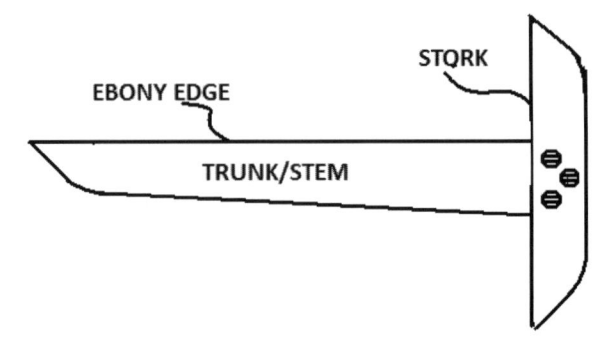

Figure 2-4: Tee square

e. *Drawing pencils / clutch pencils*: Pencils are either clutch or wooden types. Pencil lead is available in different hardness numbers (6B-6H). The person who is tracing has to find the hardness suitable - that which gives even, black lines without leaving loose graphite which will blacken the drawing.

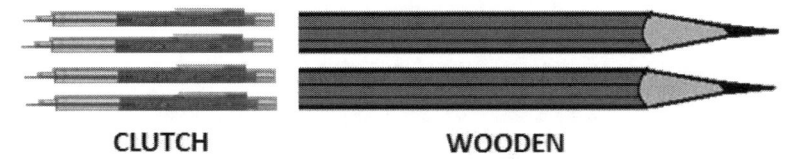

CLUTCH   **WOODEN**

Figure 2-5: Pencil types

Pencil grades – Usually pencils are H, B, HB or F series depending on the hardness, boldness or a combination of both.

i. *Hard Pencils*: The hard lead pencils are used for construction line on technical drawings. A hard pencil is used in producing construction lines because it is difficult to produce a very faint line with a soft pencil

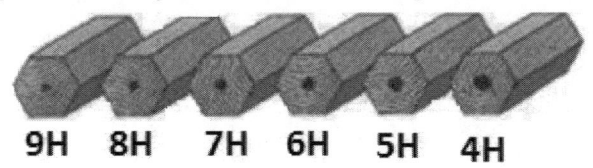

9H   8H   7H   6H   5H   4H

Figure 2-6: Hard pencil

ii. *Medium hard pencil*: They are used for general purposes I drawings. The harder grades are for instrument drawings and the softer are for sketching.

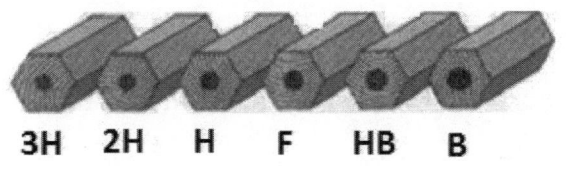

3H   2H   H   F   HB   B

Figure 2-7: Medium hard pencil

iii. *Soft pencils*: Soft lead pencils are used for technical sketching and artworks, but are too soft for instrument drawings. Soft; black lines have a tendency to smudge. The tip also wears down quickly and need constant re-sharpening. This is not so with a hard pencil

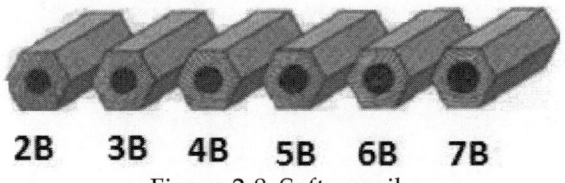

2B   3B   4B   5B   6B   7B

Figure 2-8: Soft pencil

f. *Pencils sharpeners*: Pencil sharpeners could either be razor blade, nail file or smooth sand papers. They are used in sharpening pencil and shaping the tip (pencil lead) to required shape and size.

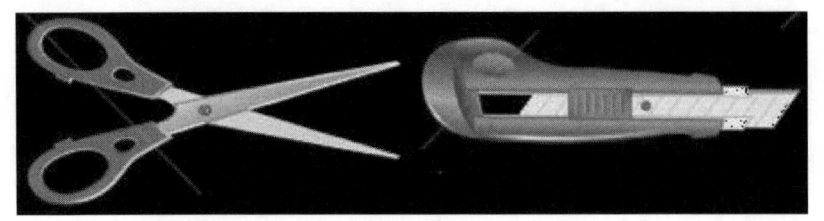

Figure 2-9: A pair of scissors and pen knife

Pencil tips:

There are two types of pencil tips obtainable in drawing exercise;

- *Chisel tip-*: Used for ruling straight lines or uniform thickness line
- *Conical tip:* Used for setting-off lengths, drawing curves and free hand sketches.

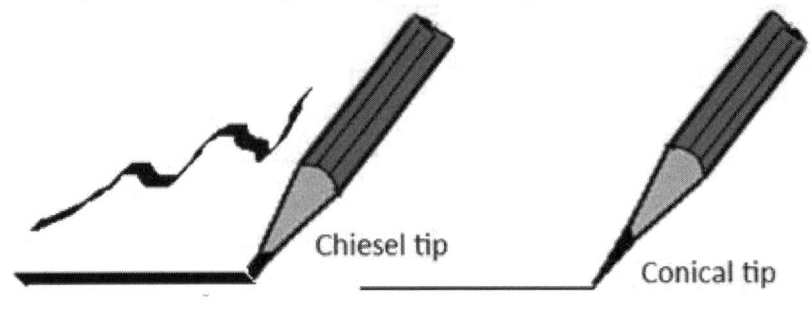

Figure 2-10: Pencil tips

g. *Set of squares*: These are celluloid plastic transparent triangular rulers with metric graduated edges for drawing inclined lines as well as drawing and transfer of parallel lines. There are three types of set squares; 30°x 60°, 45°x 45°and adjustable set squares.

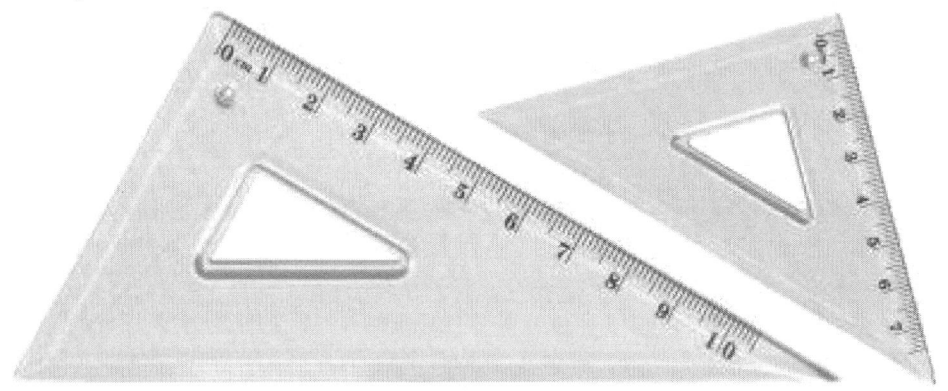

Figure 2-11: Set squares

h. *Protractor*: This is a circular or semi-circular instrument graduated in degrees which can be used to measure angles between 0 and 360 degrees. They are often made of plastic or celluloid plates.

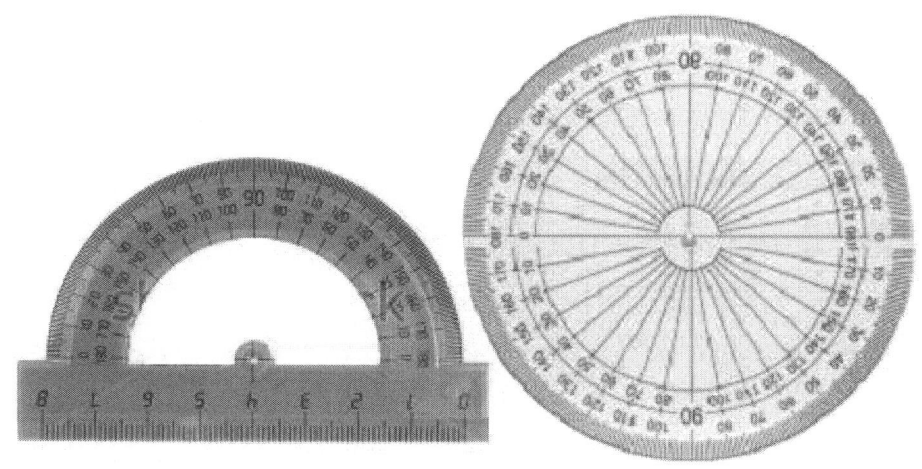

Figure 2-12: Protractors

i. *Ruler: Ruler is a long flat plastic celluloid, steel or wooden material with graduations for line or distance measurement.* If you look at your ruler, you will note that there are divisions between the cm marks and these are the mm marks. There are nine such marks allowing the lengths of say 3.1, 3.2, 3.3, ... 3.9 cm. If you use a sharp pencil (which is recommended for all drawing exercise) you should be able to measure a length to a half mm.

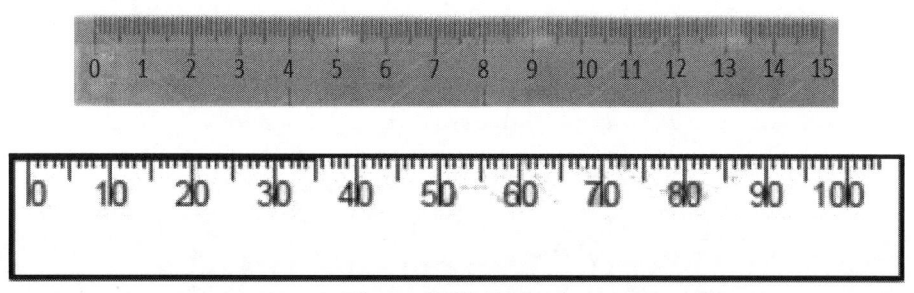

Figure 2-13: Rulers

j. *Templates*: Templates are made of plastic or celluloid materials with various shapes, letters and numerals for both lead and ink drawing and for different thickness of lines and for various uses, i.e., lettering, circles, curves, symbols, etc.

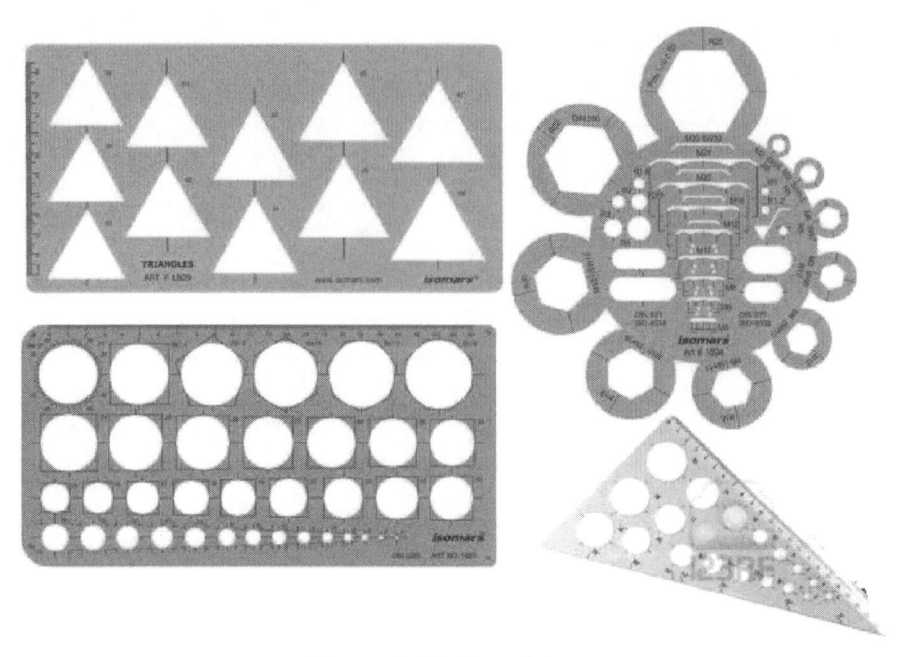

Figure 2-14: Templates

k. *Drawing set*: the drawing set is a rectangular box containing; a pair of compasses for producing circles and arcs and a pair of dividers for transferring measurements etc. Other materials include the lengthening bar, brush holder, etc.

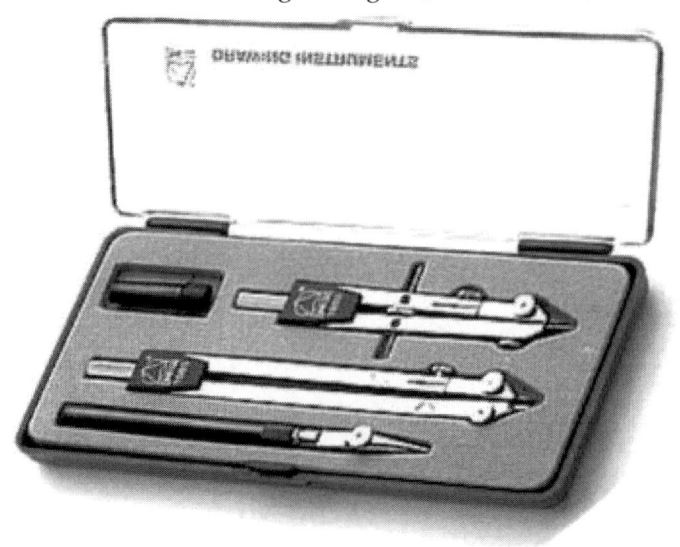

Figure 2-15: Drawing set

- *Compass*: Compass is used for drawing circles and arcs of circles. The compass has two legs hinged at one end. One of the legs has a pointed needle fitted at

30 | P a g e

the lower end where as the other end has provision for inserting pencil lead. Circles up to 120mm diameters are drawn by keeping the legs of compass straight.

Figure 2-16: Compass

- *Lengthening bar:* For drawing circles more than 150 mm radius, a lengthening bar is used. It is advisable to keep the needle end about 1mm long compared to that of pencil end so that while drawing circles, when the needle end is pressed it goes inside the drawing sheet by a small distance (approximately 1mm).
- *Dividers*: This instrument has two arms like the compass used for measurement transfer and marking out on paper. The two arms end in pins which serves as measuring

Figure 2-17: Pair of dividers

j. *Scale rules*: This is a piece of long graduated rule (either long or short) with straight edge for distance measurement and for drawing of straight line. The graduations are either in English (imperial) unit of inches and feet or metric units of millimeters, centimeters or meters. Each rule is 300mm long.

Figure 2-18: Scale rules

Scale rules are either flat edged or triangular in shape. The triangular scale rules (Rule 311-315) made from tough matt white dunirit material has six precision divided sections in every rule. The divisions common to each specific rule are as follows:

a. *Rule 311*: 1:1, 1:20, 1:25, 1:50, 1:75, and 1:125
b. *Rule 312*: 1:100, 1:200, 1:250, 1:300, 1:400, and 1:500
c. *Rule 313*: 1:2.5, 1:5, 1:10, 1:20, 1:50, and 1:100
d. *Rule 314*: 1:500, 1:1000, 1:1250, 1:1500, 1:2000, and 1:2500
e. *Rule 315*: 1:20, 1:25, 1:33 1/3, 1:50, 1:75, and 1:100

A metric scale rule is generally recommended for all drawings. Good quality scales are usually either 300mm (12") or 150mm (6") long, made of boxwood, ivory, boxwood with white celluloid edges or white or yellow plastic.

k. *Rubber eraser*: Rubber erasers are made petroleum products with ability to remove pencil or ink from paper surfaces. Erasers are used in correcting wrong impressions in your drawing. The ability to use pencil with light pressure will determine the extent to which wrong pencil impressions will be wiped off by your eraser.
l. *Fastener*: Fasteners are used in securing work piece to the drawing board. It prevents your paper from moving about thus guiding against errors in drawing. Examples include; pins, board clips or masking tape
m. *Curves* (French or German/flexible curves): French curves are templates used to draw smooth curves. These instruments are often used in producing circles and curves of fixed diameters. The French curve is fixed and hence used in fixed curves while the German curve is flexible and can be twisted to suit the desired curve.

French Curves

Flexible Curve

Figure 2-19: Curves

Standard French curves for students containing 3 shapes are made of plastic materials, transparent and are equipped with inking edge ideal for drawing smooth curves of varying radii.

n. *Tracing paper, cloth or film*: These are specially treated papers, linen and polyester films of transparent or semi-transparent nature; when placed over an original drawing, they allow the lines underneath to be clearly seen and so copied or traced. The tracings made can be used as negatives for the making of any number of further copies by the photo-printing machine.
o. *Detail paper*: This is like thick tracing paper used chiefly for preliminary sketch drawings and layouts and for final large-scale details. It is transparent enough for copying by tracing yet is white enough for original work to show up clearly.
p. *Drawing pen* for printing works etc.: Straight lines in ink are ruled in conjunction with the T-square and set-squares by means of special drawing pens such as pelican, Graphos or Rotring variants.

Figure 2-20: Drawing pen

These pens are based on fountain pen principles. Interchangeable nibs or drawing elements are used for different thickness of lines usually ranging from 0.2mm to 1.2mm. These pens should be held perfectly upright against the edge of T-square or set-square and should be drawn smoothly with even pressure from left to right or in an upward direction.

It is recommended in International Standards for drawing pens to be manufactured for the following line widths: 0.13, 0.18, 0.25, 0.35, 0.5, 0.7, 1.0, 1.4 and 2.0mm. It is preferred, for reasons of clearness, that thick lines are made twice as wide as thin lines.

q. *Drawing ink*: Black waterproof drawing ink; cleaning eraser; sharp knife or scalpel.

2. *Drafting machine for technical drawing (drafter)*

In a drafting machine, the uses and advantages of T-square, set square, scales, and protractors are combined. One end of the drafting machine (Drafter) is clamped to the left top end of the drawing board by a screw provided in the drafter (Figure 2-21). An adjustable head with a protractor is fitted at the other end of the drafter. Two blades made of transparent celluloid material are fitted to the adjustable head and are perfectly perpendicular to each other.

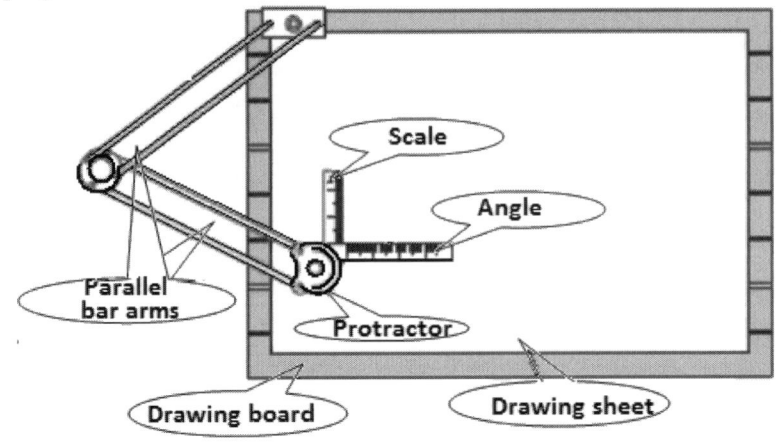

Figure 2-21: Features of drafting machine

These blades are used to draw parallel, horizontal, vertical and inclined lines. The blades always move parallel to the edges of the board. Use of drafting machine helps in reducing the time required to prepare drawing.

The drafter slides the square up, down or across on the tabletop, as required by a design, always keeping the flat side of the square flush with the tabletop's edge. This action keeps the square's blade parallel to the tabletop's edges. The drafter draws a horizontal line by resting the pencil's tip atop the square's blade, then pulling the pen right while maintaining a slight pressure

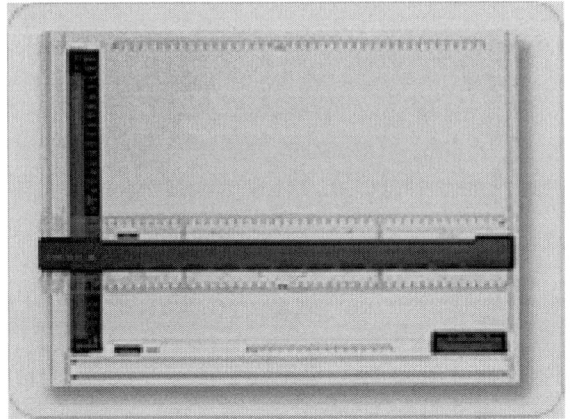

Figure 2-22: Steadler drawing board with drafter

The A3 drawing board has a size of 420 mm x 297 mm and a drafting arm with double locking mechanism, twin rail and stop and go colour indicators with variomatic drafting head made out of sturdy, break-resistant plastic, with non-slip rubber feet.

2.6 Using the instruments

Setting out drawing paper

Setting out of paper involves the preliminary steps taken in preparing for drawing practice and include the followings:

1. *Step one: Drawing board positioning*: Position the drawing board conveniently on the table with the left hand side to ensure accurate use of the T-square. Make sure that your table is cleared of any object or other tables that can cause obstruction.
2. *Step two: Drawing paper setting*: The drawing paper should be squared using the T-square to ensure that the top edge of the paper is parallel to that of the board. Place the drawing paper on the board (make sure your backing sheet is placed underneath the drawing sheet. Setting the drawing paper is done with the aid of the T-square and on no account should it be done with free (unaided) hand. This is to ensure accurate drawing of parallel lines and right positioning of drawn objects.

    Place T-square on top of the paper which can be slip/pull up or down the paper surface ensuring that the stock of the T-square is firmly in contact with the left edge of the table whenever the is to be moved. Set the top edge of the paper parallel to the edge of the T-square. Use your right hand to hold the paper down, while you slid down the T-square with the left hand. Use a fastener (masking tape) to hold the four edges of the paper firmly to the board.

3. *Step three: Drawing the border line*: The border line in every engineering drawing is a boundary line drawn to indicate the area within which objects should be drawn. Any drawing beyond this line is not reckoned with. Thus work area is demarcated by drawing the border line. Measure between 10-20mm gap from the four ends of the paper and with the combination of the T-square and set squares draw a rectangular boundary lines.

Pencils use instructions

a. Use a 0.5 mm pencil with 6H lead for drawing guidelines and construction lines.
b. Keep a .5mm pencil with H lead on hand to darken center and hidden lines. You'll also use this pencil for lettering purposes.
c. Darken thick object lines with a 0.9 mm pencil loaded with H or 2H lead.

Using the drafting tools

Drafting board: In using a drafting board, follow these procedures:

a. Lay your paper on your drafting board.
b. Line up one of the horizontal lines on your drawing with the edge of the horizontal slider on your board.
c. Tape all four corners of the paper down so that it won't wiggle or move while you draw.

The mm measurement on drafting pencils tells what the diameter of the pencil lead is. The H measurement tells how hard the lead is. For example, 6H lead is very hard, while 2H lead is much softer and H lead is the softest of all.

Using the divider: In using the dividers follow these procedures:

a. Align each arm of the dividers so that one point is laying on the start point of the measurement you want to transfer and the other divider point is laying on the endpoint of that same measurement.
b. Lift the dividers off the measurement you intend to transfer, being careful not to change their alignment.
c. Place the dividers over the location you'd like to transfer the measurement to, and make a pencil mark to indicate where each of the dividers' pointers sits. This duplicates the measurement.

Using the set of squares: Figures 3-23 and 2-24 below illustrates the use of a combination of the sets of squares to draw parallel lines

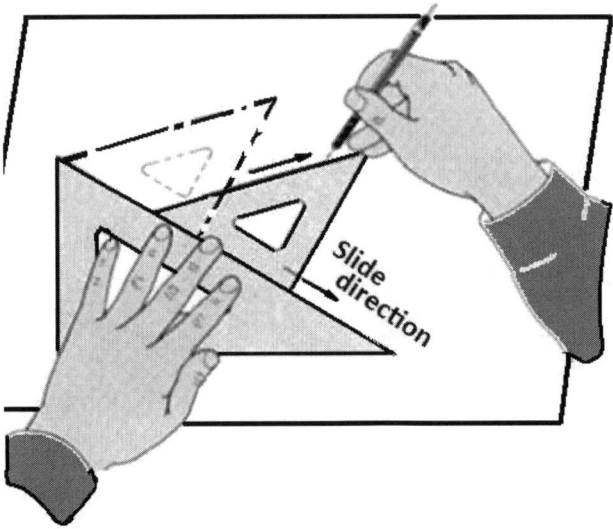

Figure 2-23: Using 45° x 60° set combination

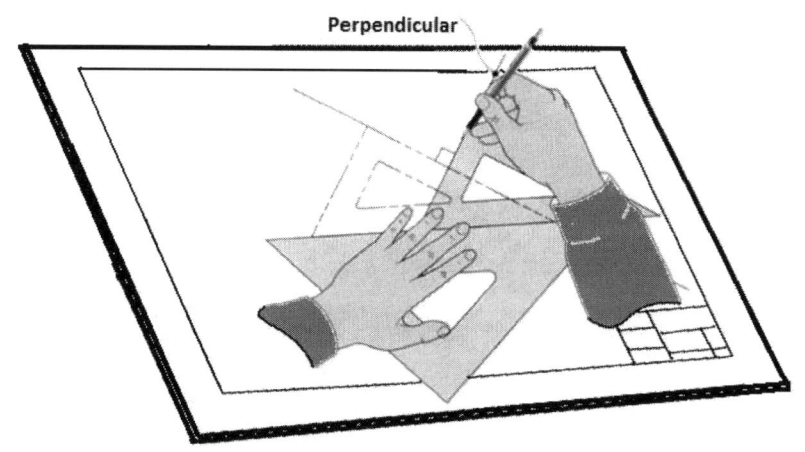

Figure 2-24: Using 60° x 45° set combination

Using the Tee-square

The head of the square, the cross member, is placed along the left edge or the top of the drafting table, while the square's blade is laid across the table's top, over the drawing paper. The draftsman slides the square up, down or across on the tabletop, as required by a design, always keeping the flat side of the square flush with the tabletop's edge. This action keeps the square's blade parallel to the tabletop's edges. The draftsman draws a horizontal line by resting the pencil's tip atop the square's blade, then pulling the pen right while maintaining a slight pressure

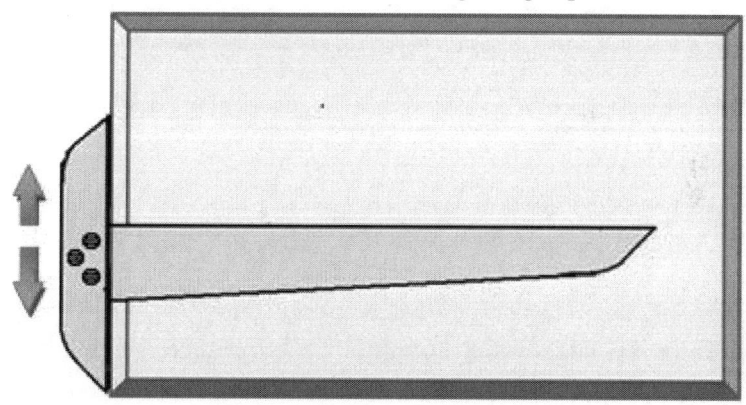

Figure 2-25: Drawing board

Using the ruler

To use the ruler and pencil to make a thick mark while you read the scale; proceed by drawing as you move away from the edge of the instrument, not drawing towards the instrument. Also, remember that you need to allow for the thickness of the line when drawing a line. When drawing a line hold the ruler using your fingers spread out along a length of the ruler, so that it doesn't slip.

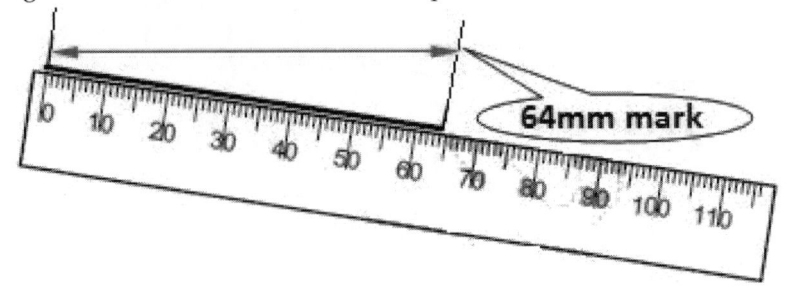

Figure 2-26: Long ruler measurement

Using the protractor

Assume you are to draw a triangle shown in Figure 2-27 accurately to scale using the protractor, proceed as follows:

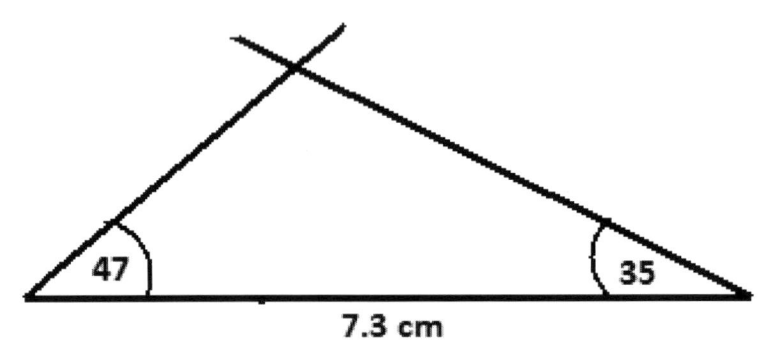

Figure 2-27: Triangle to be constructed

Step 1: First, you need to draw a base line that is 7.3 cm long with your ruler/Te-square. Leave space above it to allow the construction lines to be drawn in.

Step 2: Use a protractor to measure the angle at the left of 47 degrees using the inside scale of the protractor. Draw a line connecting the measured angle and produce the line long enough

Step 3: Move the protractor to the right and use the outside scale to mark off 35 degrees.

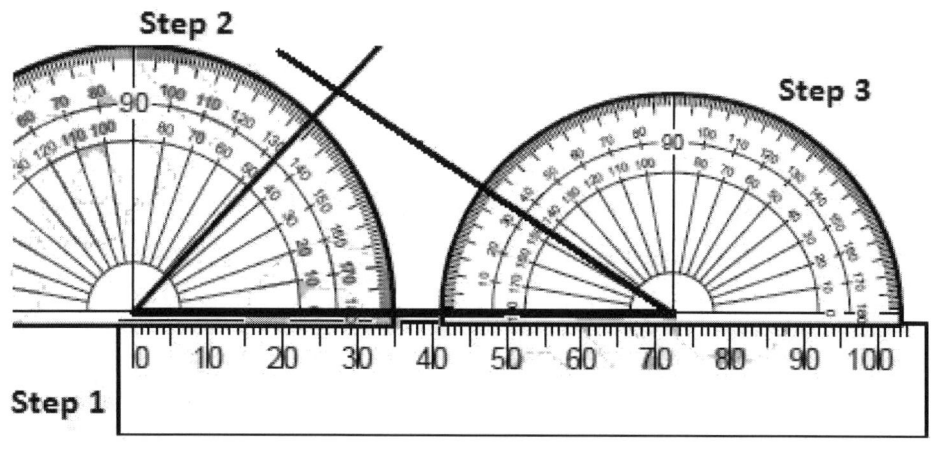

Figure 2-28: Drawing triangle with protractor

Please note that all construction lines are important part of the drawing and should **not** be rubbed out. Finally, measure the lengths and angles as accurate as possible.

Scale rule practice

Different scales are in use for different professions requiring measurement of values or quantity. Some common examples include

1. Fractional inch scale

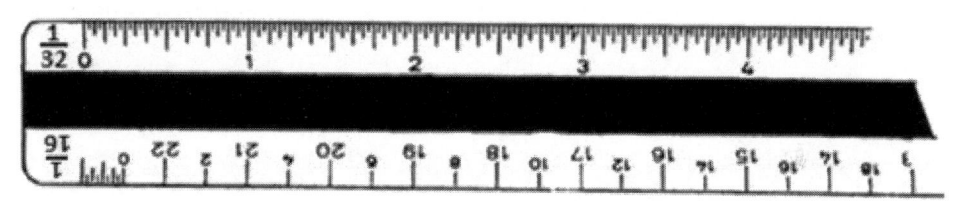

Figure 2-29: Fractional scale rule

2. Decimal inch scale

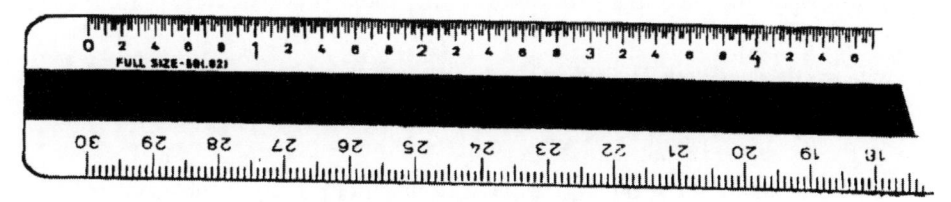

Figure 2-30: Decimal scale rule

3. Mechanical scale

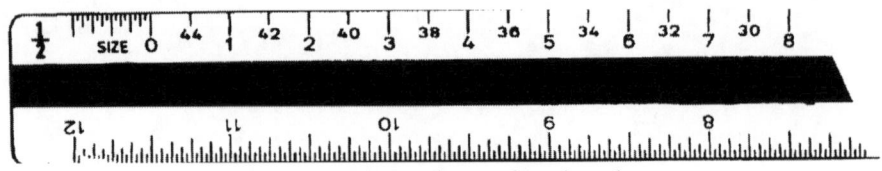

Figure 2-31: Mechanical scale rule

4. Architects scale

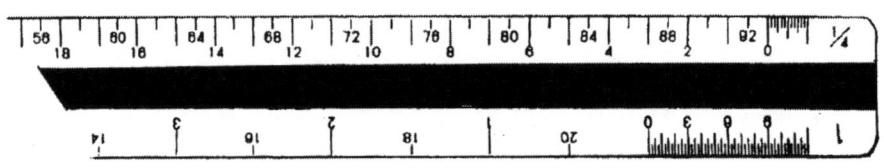

Figure 2-32: Architect scale rule

5. Engineers scale measured at a scale of 1"=10ft

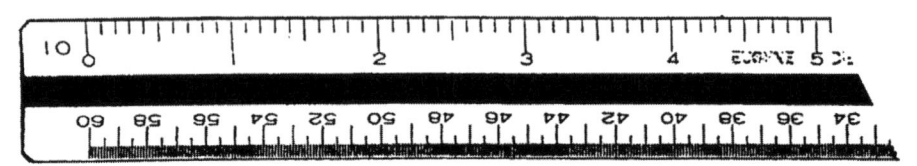

Figure 2-33: Engineers scale rule

6. Metric scale measured in millimeters

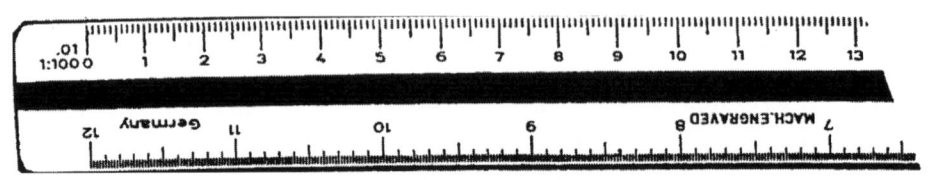

Figure 2-34: Metric scale rule

Using the protractor: In using the protractor follow these procedures:

a. Draw a straight pencil line that forms one side of the angle you need to draw. Orient the straight edge of the protractor along this line with the middle point where the vertex of the angle will be.
b. Make a small mark at the desired angle measurement on the protractor.
c. Use a straight edge--or the straight edge of the protractor--to draw a straight line from the vertex of the angle toward the mark you just made. You don't have to

connect the vertex and the mark if you don't want a long line--just make sure that, if you were to keep drawing the line, it would intersect the mark you made.

Using the curves

Before using this instrument, a draftsman will produce, using other instruments such as a compass, a series of points that are to form a curve, such as a circle or portion of a circle. French curve is equally used to connect these points with a smoothly flowing curve.

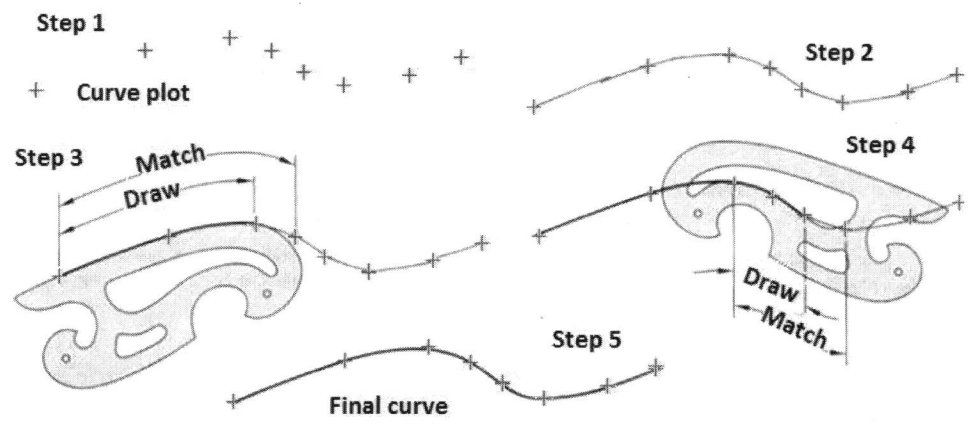

Figure 2-35: Using the French curve

To do this, lay the instrument on the paper next to the points to be connected then attempt to pass the instrument's edge through each of the points. If the instrument doesn't initially fit the points, rotate the instrument until it does. After completing this fitting process, connect the points by resting her pencil tip atop the instrument, then pulling the pencil from the leftmost point to the rightmost.

Using the drafting scale: In using the drafting scale follow these procedures:

a. Place the edge of the scale parallel to the line being measured.
b. Face the edge of the scale that you're reading toward your non-dominant side (if it's oriented vertically) or away from you (if it's oriented horizontally). This helps keep you from casting shadows on the relevant face of the scale as you work.
c. Make light marks to indicate the distance you're measuring or drawing out, as measured by the scale.
d. Adjust dividers with the scale by making a pencil line as long as the dividers should be wide, using the scale as a guide. Then adjust the dividers by orienting the points on the ends of the pencil line. Adjusting the dividers by placing the

points directly on the scale might nick the surface of the scale, making it hard to read.

Using the drafting triangle: In using the drafting triangle follow these procedures:

a. Select the vertex of the drafting triangle that has the angle you'd like to draw. For example, if you're drawing a 30-degree angle, select the vertex of the 30-60-90 triangle that measures 30 degrees.
b. Place the point of the vertex you selected where you want the vertex of the drawn angle to be. If one ray (line) of the angle is already drawn, line one of the triangle's edges up with it. If no lines have been drawn yet, just make sure the edges of the triangle are oriented as you'd like the edges of the angle to be.
c. Draw in one or both rays (lines) of the angle as necessary, extending the lines as far as required to represent the object you're drawing.

Using the drafting compass: In using the drafting compass follow these procedures:

a. Place the point of the drafting compass at the center point of the circle you intend to draw. If you're drawing an arc, imagine that the arc extends all the way around into a circle and place the point of the compass at the center of that imaginary circle.
b. Adjust the leaded end of the compass so that it touches where you'd like the edge of the arc-or circle-to be. If you're drawing an arc at a specific distance from the center point, make a line of the desired distance, adjust the point and leaded end of the compass against the ends of that line, and then place the point of the compass back at the center point of your circle or arc.
c. Grasp the middle of the compass between your thumb and fingers. Twist your fingers, applying light downward pressure on the compass to mark out the desired length of arc or circle with the leaded end of the compass.

4. *Step four: Drawing the title block*: All engineering drawings should feature an information box. This is a rectangular block that contains all necessary information about the drawing and the designer. It has no definite format and thus various forms have been used to represent the title block.

   It is conventionally placed at the lower right corner of the drawing paper. All information on the title block should be represented in uppercase form i.e. capital letters.

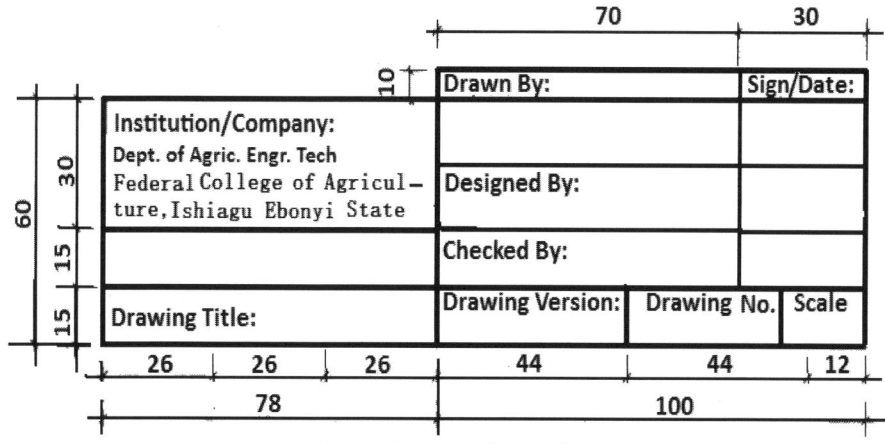

Figure 2-36: Title block

Title block information

The common information recorded on an engineering drawing work includes:

Drawing title: The title of the drawing must be specific and descriptive. For instance "ORTHOGRAPHIC PROJECTION OF A BEARING BLOCK". If all questions on the sheet are similar a general title can be given under the 'Drawing title' as indicated in the title block. The heading at the top of each page will give a guide for selection of titles.

Name of designer: The name of the person who produced the drawing. This is important for quality control so that problems with the drawing can be traced back to their origin.

Name of checker: In many engineering firms, drawings are checked by a second person before they are sent to manufacturer, so that any potential problems can be identified early.

Drawing version: Many drawings will get amended over the period of the parts life. Giving each drawing a version number helps people identify if they are using the most recent version of the drawing.

Date: The date the drawing was created or amended on.

Scale: The scale of the drawing. Large parts won't fit on paper so the scale provides a quick guide to the final size of the product.

Projection system: The projection system used to create the drawing should be identified to help people read the drawing. (Projection systems will be covered later).

Company/institution name: Many CAD drawings may be distributed outside the company so the company name is usually added to identify the source.

Other drawing details

Numbering question: All drawing questions answered on each page of paper should be numbered. This number, placed in a circle of 10mm diameter, will give identification in the same style adopted throughout the book. If mixed questions from different pages are used on one sheet, alternative method may be adopted.

Drawing accuracy: Accuracy in drawing depends on the effective use of a hard pencil, 2H or 3H, sharpened to long point. A small file such as nail file or smooth sand paper is best for keeping the point sharp. Clean the file immediately after sharpening by tapping on the end of the bench (on no account should the lead be grinded on the table top or studio wall and floors).

5. *Step five: Drawing layout*: It is important that you follow some simple rules when producing an engineering drawing which although may not be useful now, will be useful when working in industry. To produce a tidy appearance working on large sheets of paper, setting out small questions of varying sizes presents a problem of organization of material, and some guidance should be given. A2 paper (594x240) is the normal size of paper used for this type of class. A line drawn across the middle of the paper reduces the area and helps to keep small questions in better order.

2.7 Drawing reproduction

Prints of the original drawings are always used to present the project to the client, government authorities, manufacturers, building contractors, etc. In practically all cases, one of the following processes will be used:

Electro-static copying, used in most modern photocopying machines, has the advantage that the original may be on opaque paper. But most machines have a maximum size of A4 and even very expensive machines will not go beyond the A3-size.

Where there are no machines available, copies can be made by exposing the sensitized paper overlaid with the translucent original to sunlight for a few minutes and then developing the copy with ammonia.

Printing colours: Prints are available in three colours: *black* for architectural drawings, *blue* for design drawings and *red* for installation drawings. When drawings are submitted for printing, they should be rolled with the side carrying the text outwards, otherwise they may make a roll inside the printing machine and be destroyed.

Printing care: Due to shrinking or the method of copying, prints are seldom absolutely to scale. Accordingly, one should never obtain dimensions by measuring on a construction drawing, with a scale on the print! Original drawings should be stored unfolded either hanging or lying on shelves or in drawers. A simple hanger can be made from a piece of cardboard with two clothes pegs glued to the surface as shown in Figure 2-36.

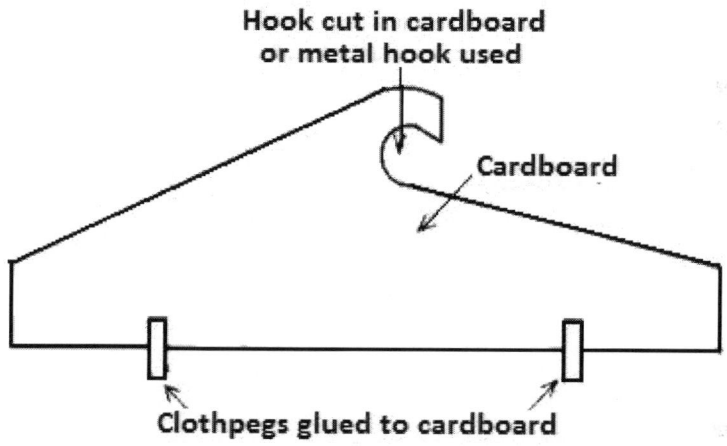

Figure 2-37: Simple cardboard hangers for drawings

Printing storage: The drawings should be stored in a cool, dry and dark room. It is well to note that a large stack of drawings can be very heavy and put a considerable load on shelves, drawers and hanger rails. Dust can be a problem in the dry season and if shelves are used, measures for control of termite and insect attacks may be necessary. Copies can be stored in the same way as originals or, in addition, folded in binders or rolled. They should be stored in darkness to avoid fading.

2.8 Scales in drawing

Scaling is used to depict objects on paper that are either larger or smaller than the paper. Scales in drawing is a measure or ratio of drawing to object as expressed below:

$$Scale = \frac{Drwaing\ size\ on\ paper}{Object\ size\ on\ paper/field} \quad \ldots\ldots\ldots\ldots 2.1$$

For instance, the an object size 20mm represented on a final drawing size of 1mm is represented as 1mm to 20mm (1:20) and interpreted as 1mm size on the drawing represents 20mm size of the object.

Representative fraction (RF)

The representative fraction of a scale is depicted as the ratio of the drawing size on paper to the ratio of the object size (on paper) or in reality. A drawing will thus have a representative fraction, RF written as

$$Representative\ Fraction, RF = \frac{Drwaing\ size}{Object\ size} \quad \ldots\ldots\ldots\ldots 2.2$$

For instance, the scale represented above has a representative fraction of

$$RF = \frac{1}{20}$$

will be written as;

$$RF = 1:20$$

Scales in drawing can be a reduction scale, enlargement scale or full scale.

Reduction scale

If the object is larger than the paper, then the views of the object are *scaled down (reduction scale)*. For instance if the object drawing below is to be drawn to a scale of 1:5, this implies that the object has dimensions 5 times larger than the drawing as represented in figure 2-38.

The scale represented above has a representative fraction of

$$RF = \frac{Drwaing\ size}{Object\ size} = \frac{1}{5}$$

Each final drawing dimension is determined by multiplying the RF by it thus:

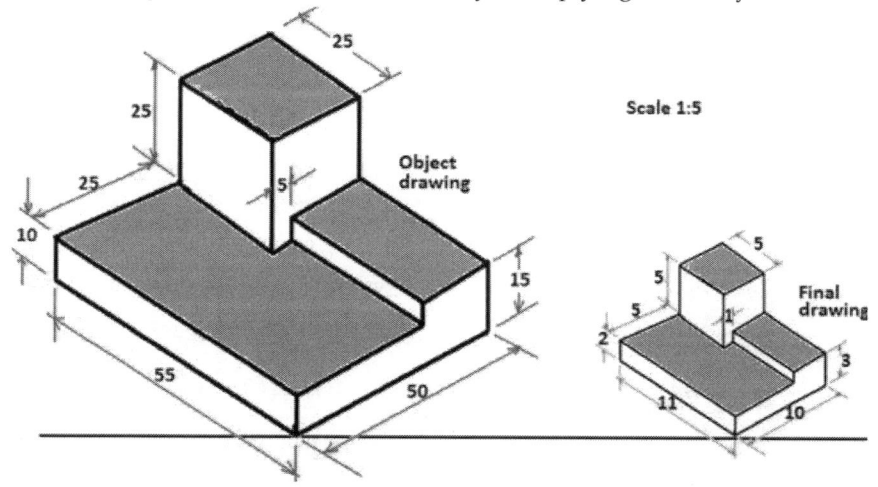

Figure 2-38: Reduction scale

Object side represented by 10mm will be $RF \times 10 = \frac{1}{5} x10 = 2mm$

Object side represented by 55mm will be $RF \times 55 = \frac{1}{5} x55 = 11mm$ in that order for all other dimensions

Enlargement scale

Enlargement scale is produced when the object is smaller than the paper some details requires expression, and then the views of the object are *scaled up (enlarged)*. For instance using the object drawing described above is to be drawn to a scale of 5:1; this implies that the object has dimensions 5 times smaller than the drawing as represented in Figure 2-39.

The scale represented above has a representative fraction of

$$RF = \frac{Drwaing\ size}{Object\ size} = \frac{5}{1}$$

Each final drawing dimension is determined by multiplying the RF by it thus:

Object side represented by 2mm will be $RF \times 2 = 5x2 = 10mm$

Object side represented by 11mm will be $RF \times 11 = 5x11 = 55mm$ in that order for all other dimensions

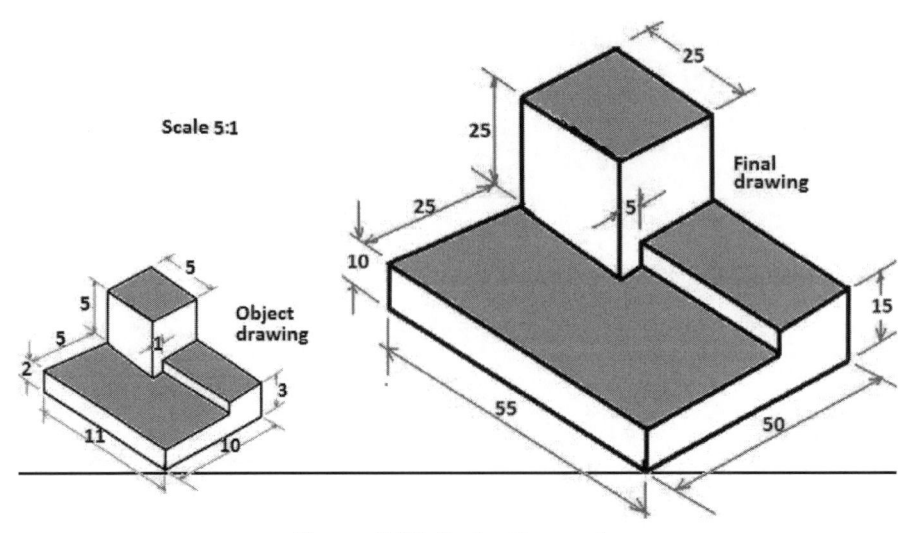

Figure 2-39: Reduction scale

Full scale

Full scale drawing is produced when the object on the paper fits or has same dimension as the drawing, then the views are depicted at *full scale*. For instance using the object drawing described in the two previous examples above is to be drawn to a scale of 1:1; this implies that the object has dimensions similar as the drawing as represented in Figure 2-40.

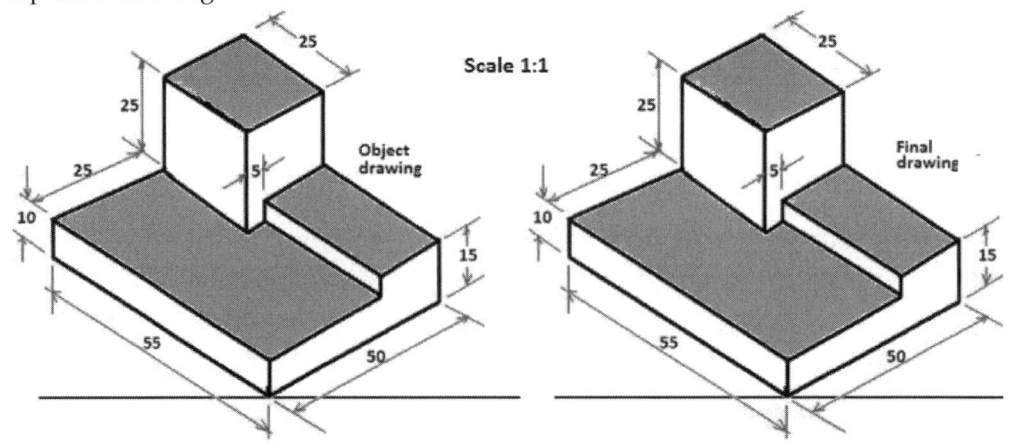

Figure 2-40: Full scale drawing

The scale represented above has a representative fraction of

$$RF = \frac{Drwaing\ size}{Object\ size} = \frac{1}{1}$$

48 | Page

Each final drawing dimension is determined by multiplying the RF by it thus:

Object side represented by 10mm will be $RF \times 10 = \frac{1}{1} \times 10 = 10mm$

Object side represented by 55mm will be $RF \times 55 = \frac{1}{1} \times 55 = 55mm$ in that order for all other dimensions

Types of length measuring instruments

The steel rule: This is often incorrectly referred to as a scale. A steel rule is the simplest of the measuring tools found in the workshop and drawing studios for linear measurements. The three basic types of rule graduations are common as shown in Figure 2-41 which include

1. Metric,
2. Fractional and
3. Decimal graduations.

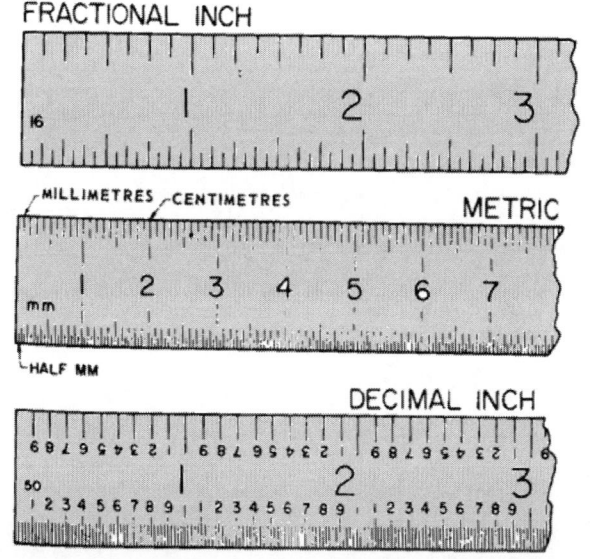

Figure 2-41: Fractional, metric and decimal graduated rules

Reading the measuring rule

A careful study of the enlarged section of the rule, Figure 2-42, will show the different fractional divisions of the inch from 1/8 to 1/64 in. The lines representing the divisions are called graduations. On many rules, every fourth graduation is numbered on the 1/32 edge, and every eighth graduation on the 1/64 edge.

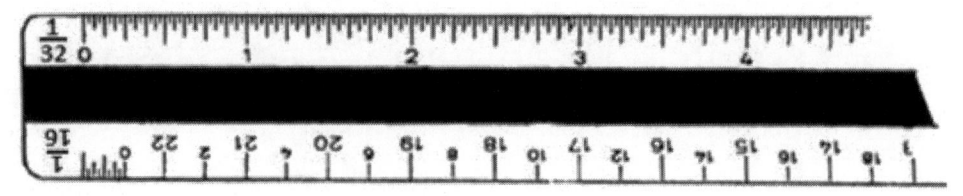

Figure 2-42: Fractional scale rule

The best way to learn to read the rule is to practice the 1/8 and 1/16 measurements and the 1/32 and 1/64 measurements until you become proficient enough to read measurements accurately and quickly. Some inch based steel rules are graduated in 10ths, 20ths, 50ths and 100ths. Fractional measurements are always reduced to the lowest terms such as a measurement of 14/16 is 7/8, 2/8 is 1/4 etc. Additional practice will be necessary to read these rules accurately and quickly.

Consider the example below, read each measurement and record your answers according to stipulated measured scale in the table provided

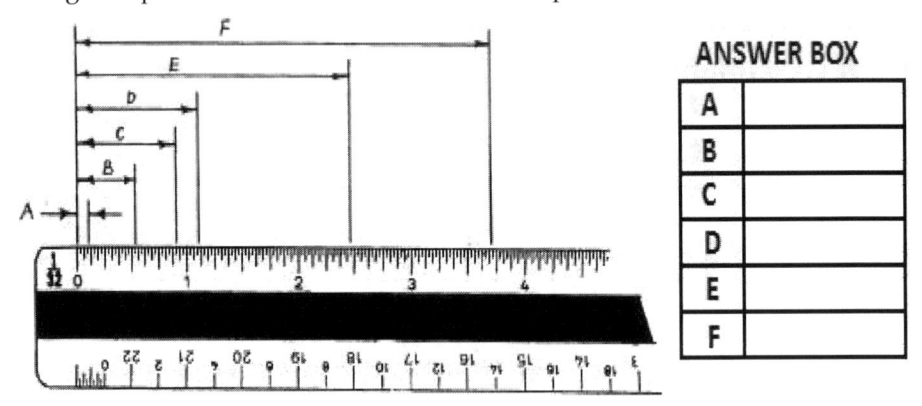

ANSWER BOX	
A	
B	
C	
D	
E	
F	

Figure 2-43: Fractional scale rule

Solution

Note that the rule is scaled 1: 32 i.e. the rule is a $\frac{1}{32}$ fractional scale. This implies that every graduation is a multiple of 2. For instance between 0 and first graduation a value of 2 is assigned, for 2nd graduation, a value of 4 is assigned until the 16th graduation (designated as 1) which is 32th mark which equals 1inch.

Taking readings from the 0 point graduation,

Reading A is $0 + \frac{1}{32}$ x number of graduations (1 in this case) x 2 in inches

$$= 0 + \frac{1}{32} \times 2 \times 2 = \frac{1}{16} \ inch$$

Reading B is $= 0 + \frac{1}{32} \times 16 = \frac{1}{2}\ inch$

Reading C is $= 0 + \frac{1}{32} \times 28 = \frac{7}{8}\ inch$

Reading D is $= 1 + \frac{1}{32} \times 2 = 1\frac{1}{16}\ inches$.......

Scales in building drawings

When the drawing of a building or object which is larger than the drawing sheet is required, it is necessary to select a suitable scale so that the final drawing and the object will be in proportion. The rule for building drawings has two lines of divisions along each edge in the proportions of 1:1, 1:100, 1:20, 1:5, 1:50, 1:250 or 1:2500. The following are the recommended scales for use with the metric system;

Detail drawing: 1:1, 1:5, 1:10, 1:20

Constructional plans, elevations and sections: 1:50, 1:100, 1:200

Layout and site plans: 1:500, 1:1250, 1:2500

All types of projections can be drawn or constructed to scale, but they become really useful to the building designer once the technique is so familiar that most of the details in the drawing including major contours in map pictures may be drawn freehand.

generally in all mechanical drawings, decimal, inch or metric scaling are commonly in use. The number on the left of the colon indicates the units on the page (drawing). The number on the right of the colon indicates the units of the object or PAGE (Drawing): OBJECT

For instance if you want to use a decimal-inch or metric scale of 1:2, it is interpreted as:

One unit on the page (drawing) = two units on the object, or *Scale* 1: 20 or may be written as $\frac{1}{2}$;

2.9 Components of drawing

Components of a complete drawing include in parts the followings

List of drawings

Where there are several drawings for a building project loss or omission of a single drawing may be avoided by listing all of them on an A4 paper. Information on latest revisions will ensure that all drawings are up to date.

Technical specifications

The technical specifications set out quality standards for materials and workmanship in respect to elements that have been described in the drawings. Where general specifications are available they are commonly referred to as *divergences* and are specified in the technical specifications.

However, in drawings for small- and medium-sized farm building projects, one tends to include much of the information normally given in the specifications, directly on the drawings.

As a basic rule, information should only be given once, either in the specifications, or on the drawing. Otherwise there is a risk that one place will be forgotten in a revision and thus cause confusion.

Functional and management instructions

Frequently, information has to be transferred to the person using a structure to enable him to utilize it in the most efficient way or the way intended by the designer. In, a pig house, for example, different types of pens are intended for pigs of certain age intervals. Alleys and door swings may have been designed to facilitate handling of pigs during transfer between pens. In a grain store the walls may have been designed to resist the pressure from grains stored in bulk to a specified depth.

Bill of quantities

The bill of quantities contains a list of all building materials required and is necessary to make a detailed cost estimate and a delivery plan. It cannot be produced however, until the detailed working drawings and specifications have been completed.

Cost estimate

The client will require a cost estimate to determine whether the machine/building should be constructed or not. He needs to know whether the proposed design is within his financial means and/or whether the returns of the intended use of the building will justify the investment.

Time schedule

The farmer may obtain information on when he and any farm labourers will be involved in the construction operations, when animals and feed should be delivered, when a breeding programme should be started or the latest starting date for the construction of a grain store to be completed before harvest. All this is the type of information needed to enable the returns of the investment to be collected as early as possible. A contractor will require a more detailed chart for the actual construction operations to promote an economical use of labour, materials and equipment.

Exercise

1. *Drafting tools*

    a. Pencils are essential drawing aids in technical drawing.

        i. What are the uses of pencil in drawing?
        ii. What are the types of pencil grades we have?
        iii. What are the application purposes for each grade of pencil?
        iv. List the two types of pencil tips?

    b. Thin lead mechanical pencils are available with four different lead diameters. What are they?
    c. List two advantages of a thin-lead mechanical pencil.

        i. _____
        ii. _____

    d. From the following list of pencils list them in order from softest to hardest. 2B, 6H, 6B, 9H, H, B, 2H ___ ___ ___ ___ ___ ___ ___
    e. When selecting the grade of pencil lead, what lead should be used for construction lines? _____
    f. For mechanical drawings on drawing paper or tracing paper, the lines should be _____ , particularly for drawings to be reproduced. The lead chosen

must be soft enough to produce _____ lines, but hard enough not to be_____ too easily.

2. Standard USA paper sizes are designated by the letters A, B, C, D, E. For each letter below write the paper size next to the letter. A_____ B_____ C_____ D_____ E_____
3. List all the common information recorded on an engineering drawing
4. Scale sheet practice:
   Read each measurement and record your answers according to stipulated measured scale in the table provided

   a. Fractional inch scale

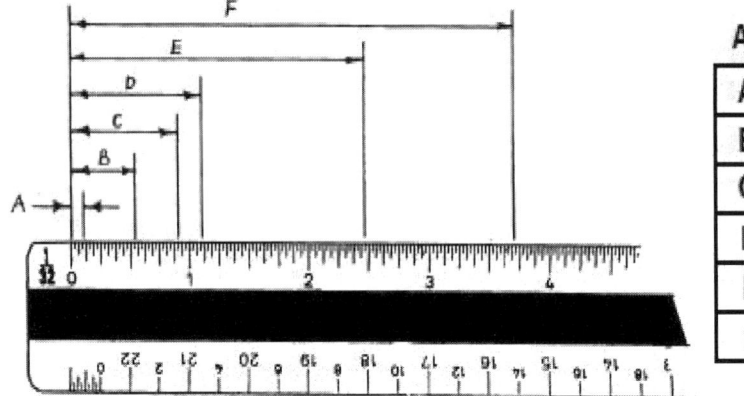

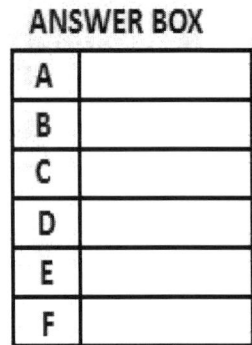

ANSWER BOX	
A	
B	
C	
D	
E	
F	

   b. Decimal inch scale

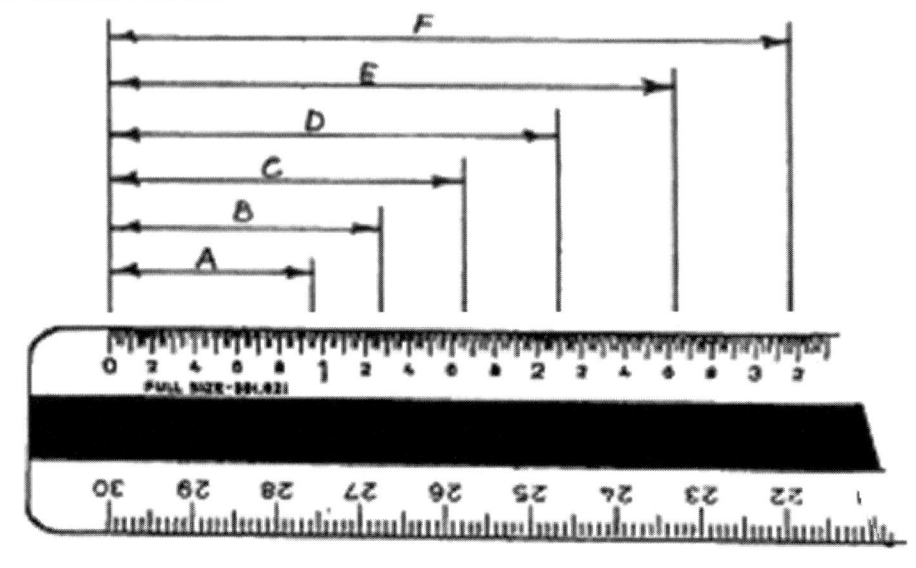

c. *Mechanical scale*

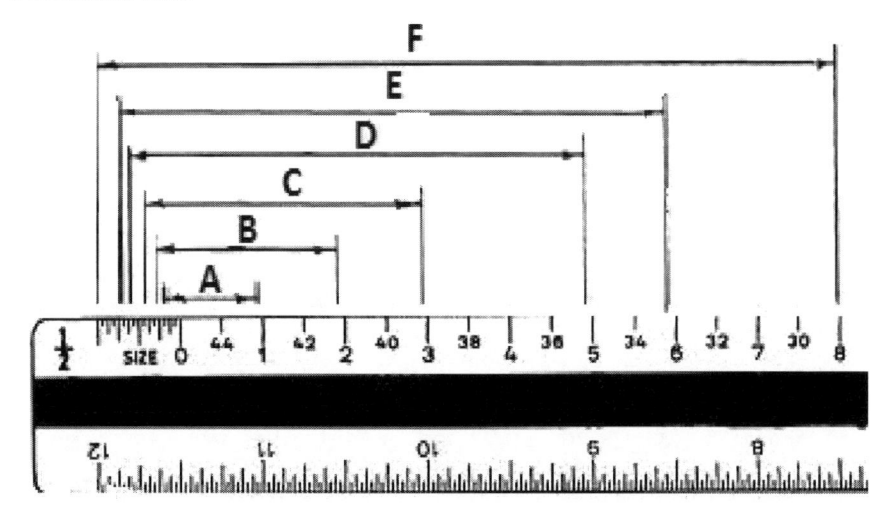

d. *Architects scale*

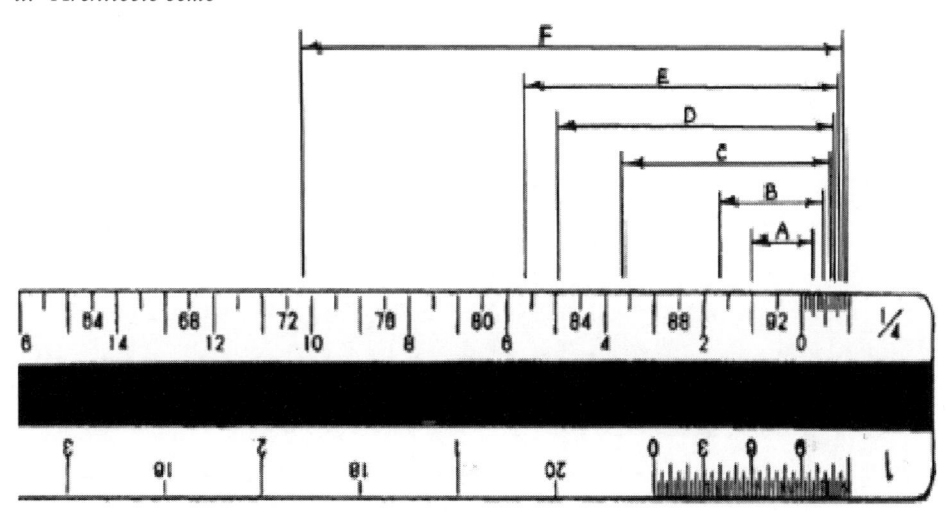

e. Engineers scale reading

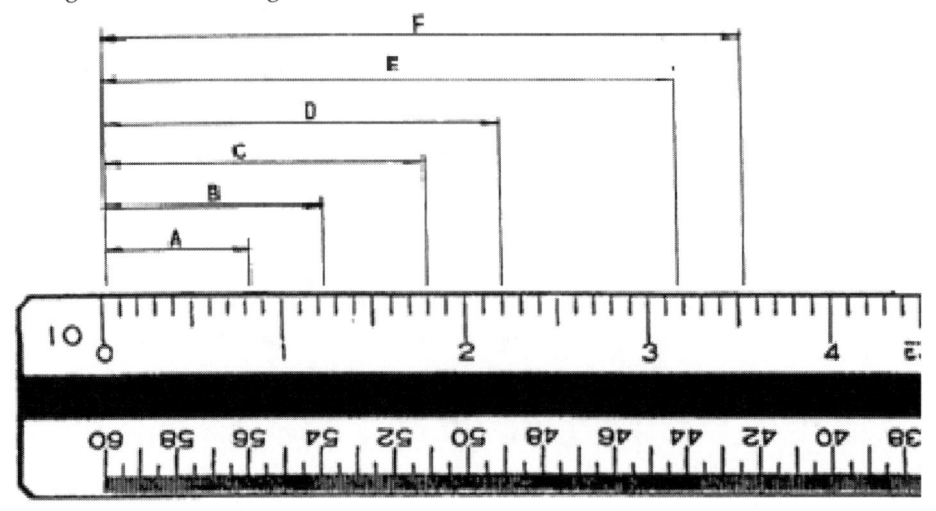

f. Metric scale reading

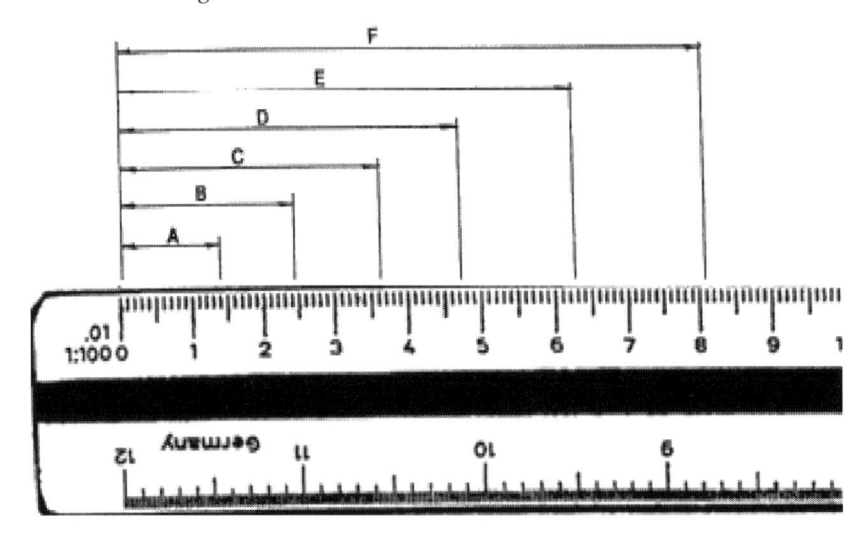

Chapter 3

Lettering Principles and Practice

3.1 Introduction

Lettering is a practice of inclusion of descriptions in forms of numbers, alphabets, notes and expressions as an integral part of any engineering drawing. For instance, every drawing needs a title, and often subtitles are required. In addition, in order to make drawing easier to understand and more useful to operators, descriptive notes and dimensions will generally be required.

Lettering practice is essential in drawing because there is a standard which must be maintained internationally in all drawing presentation, symbols and dimensioning. It is therefore necessary to discuss the importance, rules, requirements and principles of lettering in engineering drawing.

3.2 Importance of lettering

Lettering offers the following importance to our drawing:

1. Lettering helps in the identification of drawn objects.
2. It gives more clarity to drawing through titles and sub-titles and makes it easily read.
3. It gives detail description of drawing parts or components
4. Additional information are given about the origin of the drawing through lettering
5. Other necessary detail information about the designer, author, etc. is further given in lettering.
6. Orderly presentation of procedure of production of draw object is given by lettering.

3.3 Types of lettering

The following types of lettering are often encountered in drawings

1. *Upper case letters*: These are letters written in capital letters only e.g. ORDERLY
2. *Lower case letters:* These are letters written in small letters within the body or content of text as indicated in the underlined e.g. examine your protractor

3. *Superscript letters*: These are letters that are indented above a letter or figure to indicate emphasis or power e.g. $A^{xy}, 8^{10}$ etc.
4. *Subscript letters* are letters indented below another letter or word e.g. x_{yield}, δ_0 etc.

3.4 Style of lettering

Two styles of lettering are commonly used in technical drawing. Whichever one you decide to use, there is no limitation but ensure consistency of style.

1. *Straight/block style*: This style involve letters written in straight standing form
2. *Italics*: These are letters written in slanting form within the body of text to give an emphasis e.g. When measuring, it is better to take a *short faint line* rather than *a dot*.

3.5 Lettering fonts in drawing

Fonts indicate the type of lettering intended to use in writing. Various types of font are available for lettering practice, most especially in computer aided designs and drawings. Typical name of font and font style in use includes:

Arial black: **Technical drawing practice**
Vedana: Technical drawing practice
Goudy Stout: **TECHNICAL DRAWING**
Showcard Gothic: **TECHNICAL DRAWING PRACTICE**

Figure 3-1: Sample lettering fonts

3.6 Lines in lettering

Two types of line are used in lettering; lines of lettering and guide lines (Figure 3-2). The description and uses of each are discussed as follows.

Lines of lettering: Lines drawn within which lettering practice is carried out is referred to as lines of lettering. They are often drawn bold and indicates the limit of lettering

Guide lines: Lines drawn within lines of lettering to guide letters against floating is called guide line. They appear faint and are parallel to the lines of lettering.

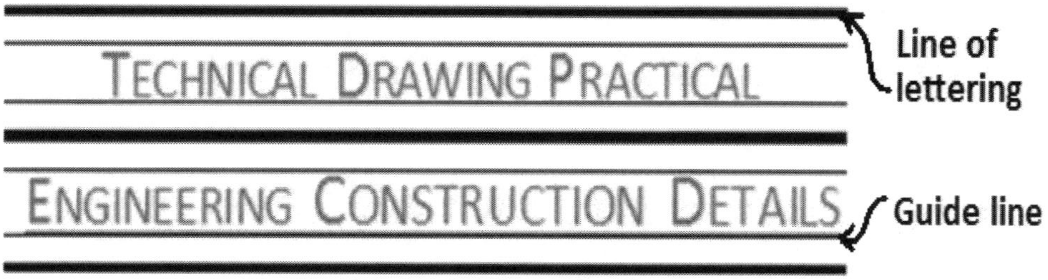

TECHNICAL DRAWING PRACTICAL — Line of lettering

ENGINEERING CONSTRUCTION DETAILS — Guide line

Figure 3-2: Lettering lines

3.7 Lettering practice

Clear lettering can be produced as easily and as swiftly as scratchy letters, by using the correct technique. Each letter character is formed by using a sequence of separate, simple strokes for the lines and bows. Use the least possible pressure and hold the pen upright and at 45° angle to the line of writing.

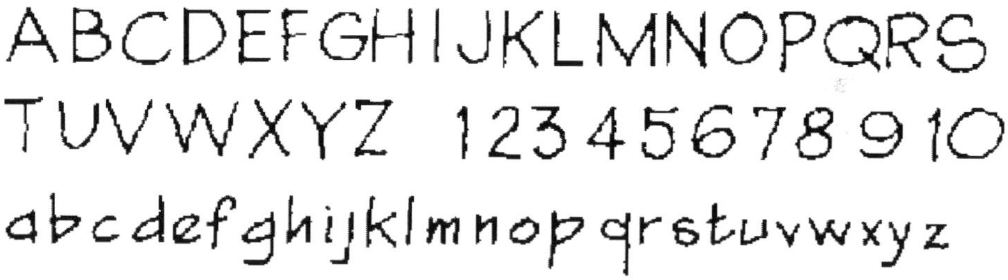

ABCDEFGHIJKLMNOPQRS
TUVWXYZ 12345678910
abcdefghijklmnopqrstuvwxyz

Figure 3-3: Samples of lettering

Lettering will normally run from left to right on the sheet and be parallel to the bottom edge. When it becomes necessary for lettering to run vertically, it should always run from the bottom upwards (this applies also to strings of dimensions).

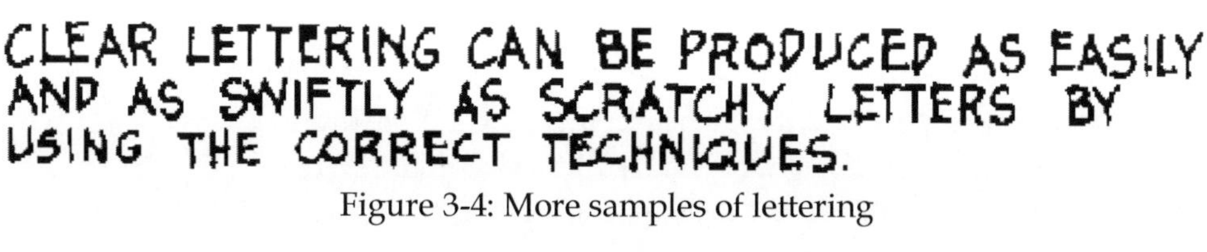

CLEAR LETTERING CAN BE PRODUCED AS EASILY AND AS SWIFTLY AS SCRATCHY LETTERS BY USING THE CORRECT TECHNIQUES.

Figure 3-4: More samples of lettering

Horizontal guidelines are essential unless the draftsman is very experienced and skillful. They may be drawn lightly in pencil for subsequent erasure when the lettering is in ink or may take the form of a closely gridded sheet laid underneath the tracing paper.

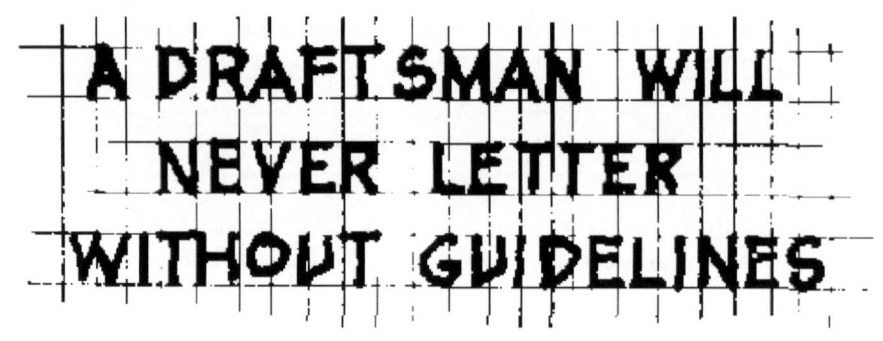

A DRAFTSMAN WILL NEVER LETTER WITHOUT GUIDELINES

Figure 3-5: Samples of lettering

Suggested heights for the letters are: 3mm for text (content) in the drawing, measurements and descriptive texts; 5 and 7mm for headings and for drawings which are going to be reduced.

Letters and words are spaced by eye rather than by measuring. If the proportion, form and spacing of the letters are properly executed, the result will be legible and pleasing to the eye.

3.8 Drawing lettering lines

The thickness of lines should be chosen so that the figures on the drawing are easy to read. For instance, the outer contour of a building drawing and the walls between rooms should be thicker than equipment, fittings and measurement lines. The major outline will then be noted first and the details later. When measuring, it is better to take a short faint line rather than a dot. The short line will disappear into the final line, but a heavy dot usually shows.

Drawing different lines

It requires practice to draw lines of even thickness and blackness with pencil. It is imperative to use a pencil with a sharp point. By rotating the pencil while drawing, the point will stay sharp for longer time. All lines should be drawn with the help of a ruler, except when sketching or drawing a perspective.

Drawing dimension lines

Dimensions are a very important part of the drawing and must be different and complete. No measurements should have to be calculated by the one who is using the drawing. Duplicate dimensions should be avoided since one may be forgotten if a change is made.

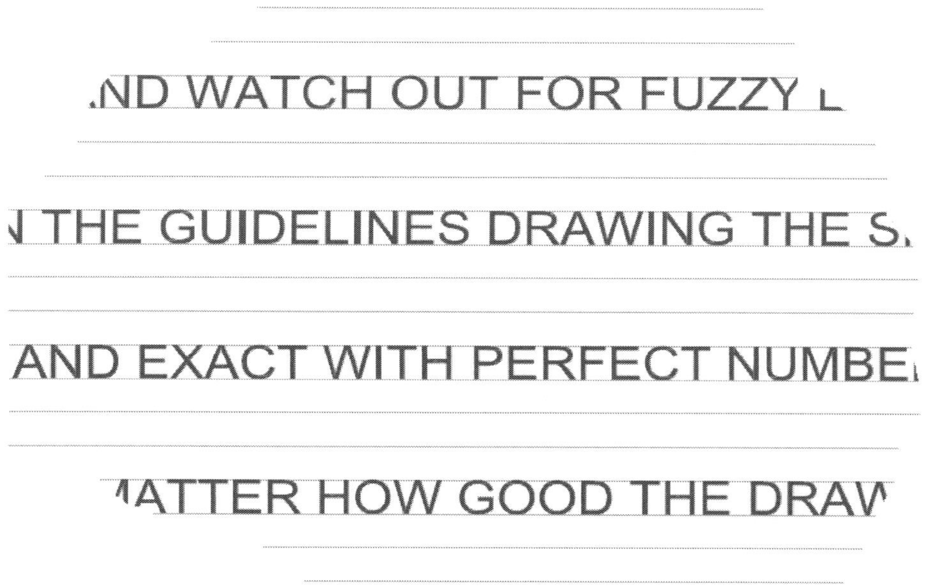

.ND WATCH OUT FOR FUZZY L

√ THE GUIDELINES DRAWING THE S.

AND EXACT WITH PERFECT NUMBE.

˄ATTER HOW GOOD THE DRAW

Figure 3-6: Samples of lines of lettering

Dimensions should be easy to read and placed where the reader will expect to find them. They should appear 1 mm above the line and be placed so that they can be read either from the bottom or the right edge of the drawing. Dimensions should appear outside the figure if it does not make interpretation difficult. Related dimensions should be placed together, preferably in the same string. *Dimensions may be given in a chain or from a common point, the latter being used mainly when surveying existing buildings.

Exercise

1. Distinguish between lines of lettering and guide lines and show this with sketches
2. Describe the two styles of lettering used in technical drawing
3. Lettering ability has little relationship to _____. You can learn to letter neatly even if you have _____ _____.
4. Distinguish between the uppercase and lower case lettering
5. Draw each letter and number in the figure below using the grid provided and following these instructions:

✓ First trace over each of the letters and numbers that are given, practice forming the correct shapes.
✓ Second draw two more of each on the empty grids to the right of each letter being as precise as you can

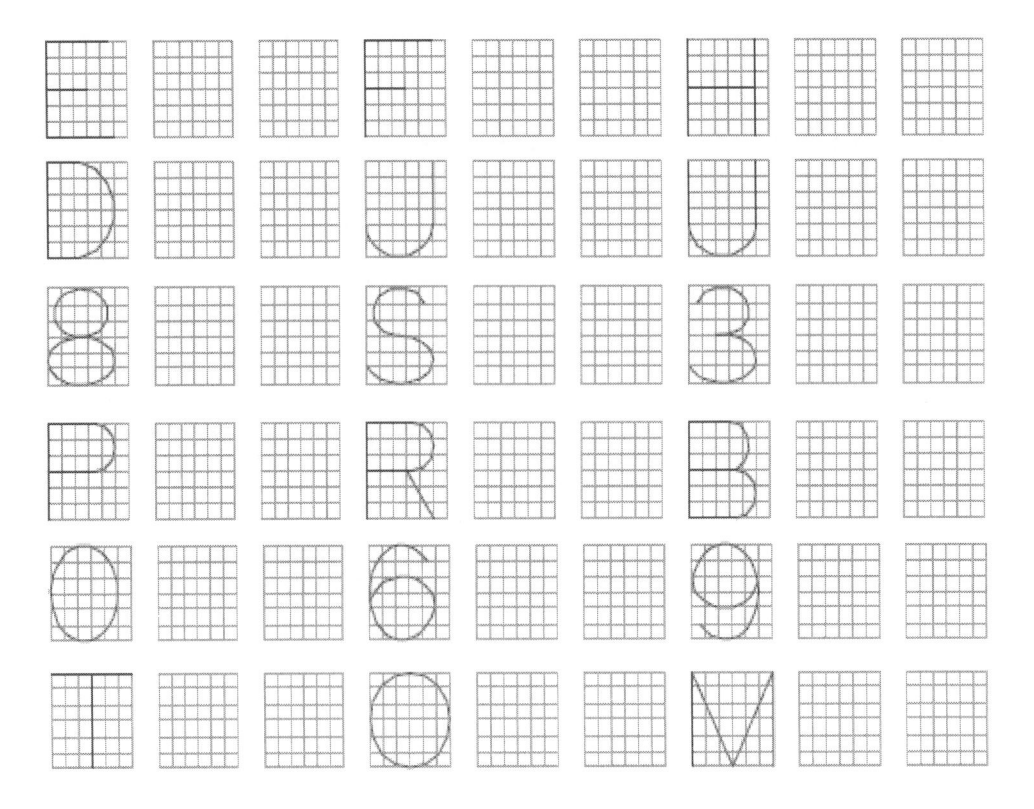

6. Produce a write up using the same font as indicated observing all lettering principles. Use 15mm for guideline
7. List six examples of lettering font and their characteristics

Chapter 4

Principles of Dimensioning

4.1 Importance of dimensioning

Dimensions are important parts of technical drawing indicating size of object. Dimension lines are drawn with the aid of projections which describe the shape and some indication of sizes. A drawing is not a complete communication unless it gives information about size in addition to the description it gives of shape. The way sizes and shapes are shown on a drawing is the technique of dimensioning. The purpose of dimensioning is to provide a clear and complete description of an object. Dimensions tell how far it is from one point on an object to another point.

4.2 Guidelines for dimensioning

A complete set of dimensions will permit only one interpretation needed to construct a part or complete object. Dimensioning should follow these guidelines.

1. *Accuracy*: Correct values must be given.
2. *Clarity and exactness*: Dimensions must be placed clearly in appropriate positions and must be exact.
3. *Completeness*: Nothing must be left out, and nothing duplicated.
4. *Readability*: The appropriate line quality must be used for legibility, letters and figures must be legible and readable. Style of writing must be constant and uniform.

4.3 Types of dimensions

There exist a number of dimension types used in mechanical drawings which include:
1. Linear dimensioning: This is the most basic type of dimensioning because it gives the straight-line distance from one point to another. This type of dimensioning is further divided into two:
   a. Coordinate dimensions
   b. Coordinate without dimension lines (Ordinate)
2. Angular dimensioning

3. Radial/diametrical dimensioning
4. Tabular dimensions dimensioning

4.4 Components of dimension

Dimensions have four basic components:

- *Dimension text*: This gives an expression or description to the part so labelled. It can either be numeric or alphabetic.
- *Dimension line and arrows*: This is the indication of the limits of description to the labelled part
- *Extension/projection lines (long or short)*: This is the projection the covers the distance between the dimension line and the actual drawing.
- *Gap/small space*: This is a small space separating the projection line from the body of the drawing. Note that the extension lines can cross over object lines (visible edges of the object) to reach their destination, but still leave a gap.
- *Notes*: Notes are added text to describe things on the drawing. The simplest type is just text, as in the title of the drawing. The other type of note contains a leader, which is an arrow that points to the subject of the note.

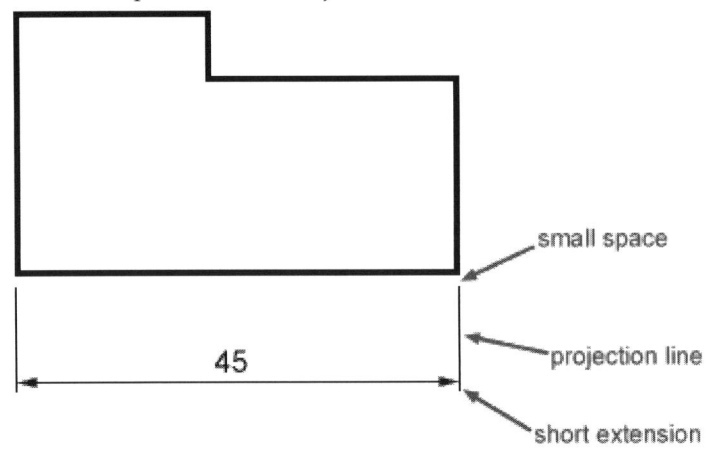

Figurre 4-1: Dimension components

4.5 Dimension placement

The placement of your dimension lines, arrows and dimension text depends on the amount of space there is between the extension lines.

a. *Dimension lines* must be arranged at suitable distance from the object, they must be in alignment, and the overall dimensions must come last. Haphazard placement of dimensions could lead to misinterpretation, hence the need for standardization. The recommendations in BS 308 covering dimensioning are extensive and need systematic teaching.

b. *Letters and figures:* Letters and figures must be placed above the line; readable from bottom or right. These are also placed on narrow spaces, an angle, circle or radii.

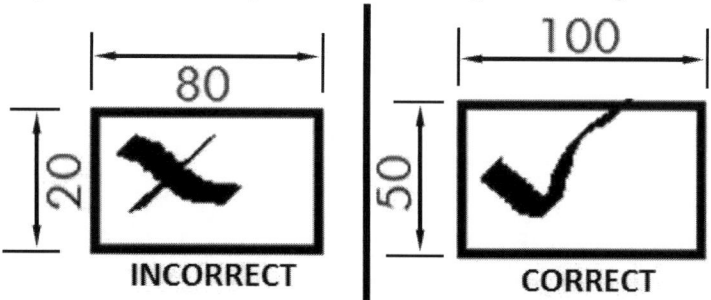

Figure 4-2: Dimension placements

The checklists of the basic knowledge of dimensioning are discussed in the following rules. As a general guideline to dimensioning, put in exactly as many dimensions as are necessary for the craftsperson to make it - no more, no less.

Placing dimensions in holes

Examples of appropriate and inappropriate placing of dimensions are shown in Figure 4-3.

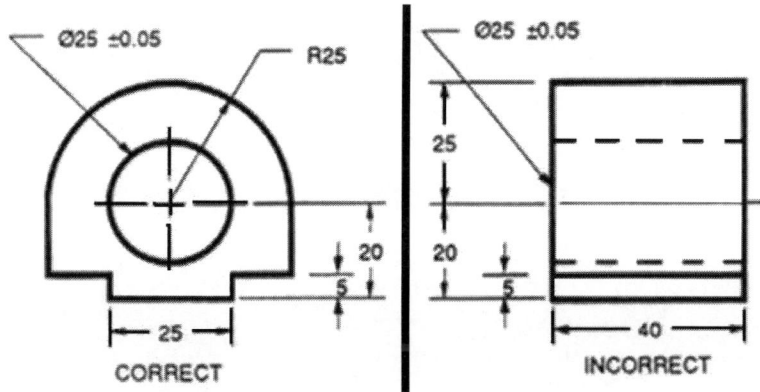

Figure 4-3: Dimension placement

In order to get the feel of what dimensioning is all about, we can start with a simple rectangular block. With this simple object, only three dimensions are needed to describe it completely (Figure 4-4). There is little choice on where to put its dimensions.

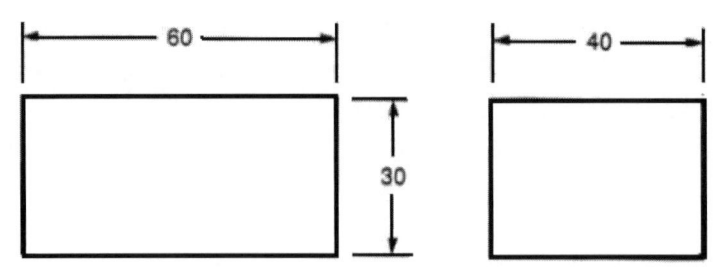

Figure 4-4: Dimension placement

We have to make some choices when we dimension a block with a notch or cutout (Figure 4-4). It is usually best to dimension from a common line or surface. This can be called the datum line of surface. This eliminates the addition of measurement or machining inaccuracies that would come from "chain" or "series" dimensioning. Notice how the dimensions originate on the datum surfaces. We chose one datum surface in (Figure 4-5), and another in (Figure 4-6). As long as we are consistent, it makes no difference. (We are just showing the top view).

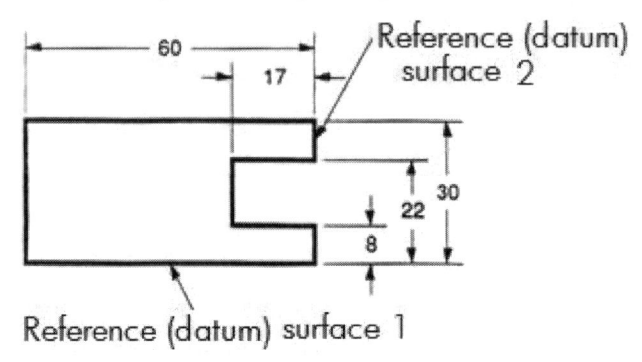

Figure 4-5: Surface datum (1)

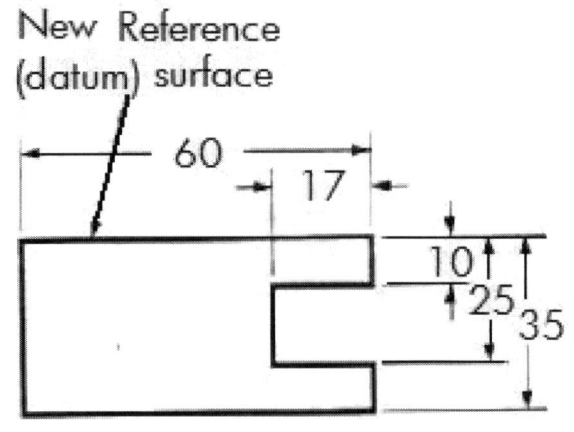

Figure 4-6: Surface datum (2)

4.6 Rules of dimension

The overriding principle of any dimensioning is *clarity*. The following rules apply to all good and best dimensioning practice.

1. *How to draw dimension lines*: Dimension lines must be medium black in strength, and there must be small gap between projection line and the object. All notes and dimensions should be clear and easy to read. In general all notes should be written in capital letters to aid legibility.

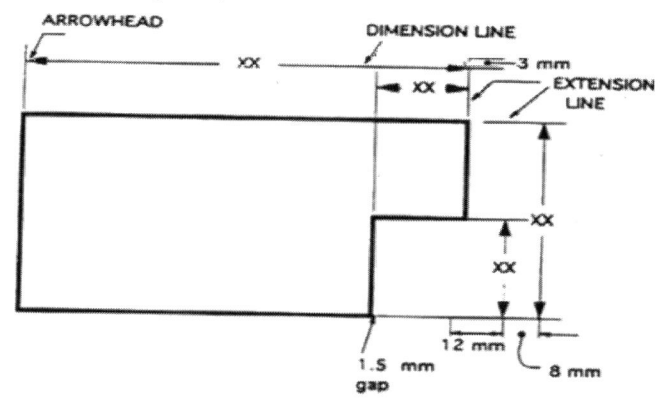

Figure 4-7: Drawing dimension lines

2. *Forming arrow head*: The arrow heads must be close to the dimension line to produce a *narrow* arrow. All arrow head sizes should have the dimension shown in the figure.

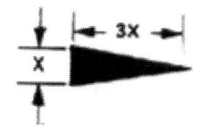

Figure 4-8: Arrowhead dimension

Samples of correctly formed and incorrectly formed arrowheads are as shown in Figure 4-9.

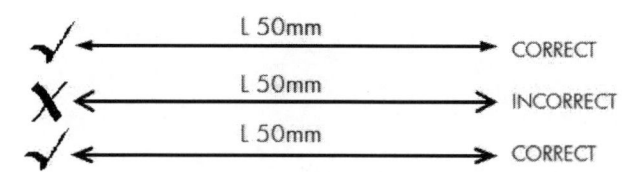

Figure 4-9: Correct and incorrectly dimensioned arrowhead

3. *Extension line/gap*: The space between the first dimension line and the object should be at least 3/8 inch (0.375). The space between all other dimension lines should be at least 1/4 inch (0.250). There should be a visible gap between the object and the origin of the extension line. Extension lines should extend 1/8 inch (0.125) beyond the last dimension line. Extension lines should be broken if they cross or are close to arrowheads.

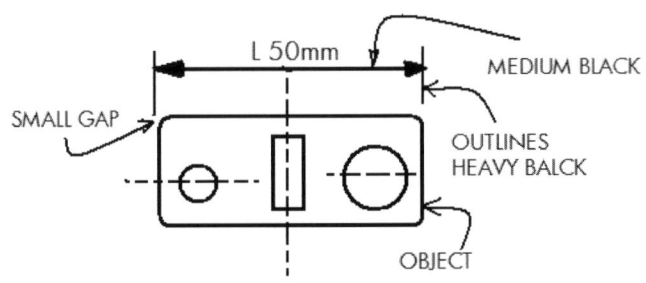

Figure 4-10: Dimension outlines

4. *Placement*: Dimensions should be placed in the most descriptive view of the feature being dimensioned. Do not put in redundant dimensions. Not only will these clutter the drawing, but if "tolerances" or accuracy levels have been included, the redundant dimensions often lead to conflicts when the tolerance allowances can be added in different ways.

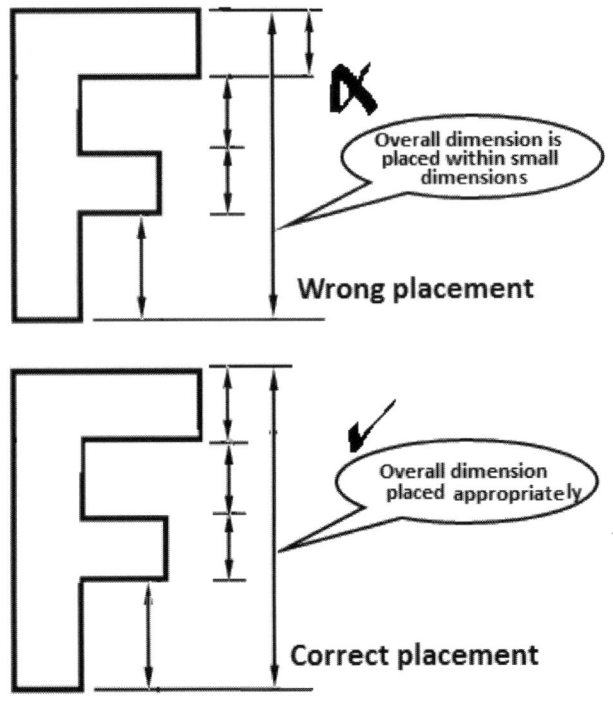

Figure 4-11: Dimension placement

5. *Lettering size*: All lettering should be of the same size and preferably not smaller than 3mm. An example typeface is shown below.

ABCDEFGHIJKLMNOPQRSTUVWXYZ
1234567890

Figure 4-12: Lettering sizes

6. *Dimension frequency*: Each feature of an object is dimensioned once and only once. Repeatedly measuring from one point to another will lead to inaccuracies. It is often better to measure from one end to various points. This gives the dimensions a reference standard. It is helpful to choose the placement of the dimension in the order in which a machinist would create the part. This convention may require some experience.
7. *Object dimension*: Objects or drawing are various shapes and sizes and therefore requires different approach. For instance
    a. *Circles*: Circle diameters are dimensioned with a numerical value preceded by the diameter symbol. Radii are dimensioned with a numerical value preceded by the radius symbol. When a dimension is given to the center of an arc or radius a small cross is shown at the center.
    b. *Cylinders:* Cylinders should be dimensioned in the rectangular view where possible. The diameter and depth of holes that are counter-bored, spot faced, or countersunk, should be specified in a note. The depth of a blind hole may be specified in a note, and is the depth of the full diameter from the surface of the object.

4.7 Types of dimensioning

1. *Parallel dimensioning*: Parallel dimensioning consists of several dimensions originating from one projection line.

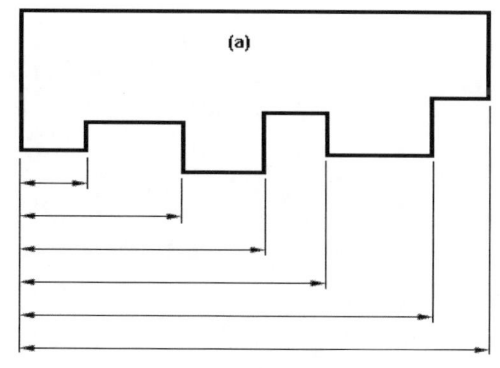

Figure 4-13: Parallel dimensioning

2. *Superimposed running dimensions:* Superimposed running dimensioning simplifies parallel dimensions in order to reduce the space used on a drawing. The common origin for the dimension lines is indicated by a small circle at the intersection of the first dimension and the projection line. In general all other dimension lines are broken. The dimension note can appear above the dimension line or in-line with the projection line

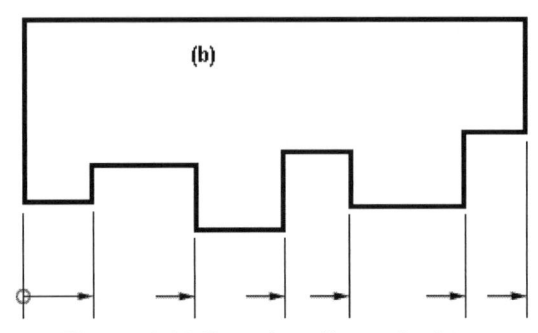

Figure 4-14: Running dimensioning

3. *Chain Dimensioning*: Chains of dimension should only be used if the function of the object won't be affected by the accumulation of the tolerances. (A tolerance is an indication of the accuracy the product has to be made to.

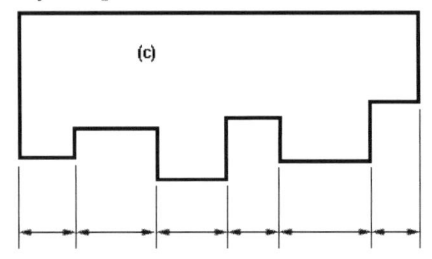

Figure 4-15: Chain dimensioning

4. *Combined dimensions:* A combined dimension uses both chain and parallel dimensioning.

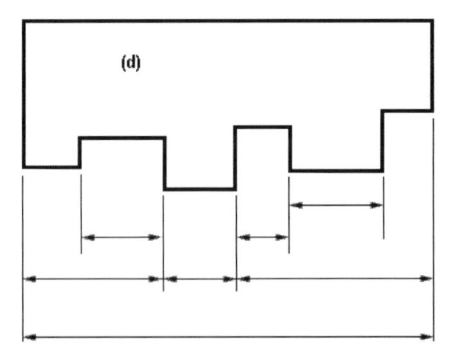

Figure 4-16: Running dimensioning

5. *Dimensioning by co-ordinates:* Two sets of superimposed running dimensions running at right angles can be used with any features which need their center points defined, such as holes.

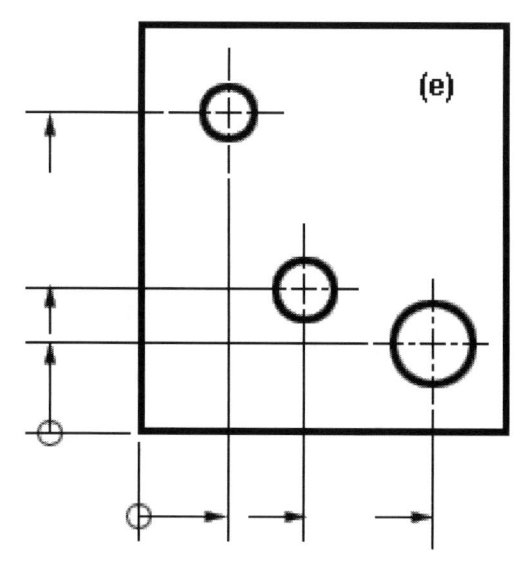

Figure 4-17: Coordinate dimensioning

6. *Simplified dimensioning by co-ordinates:* It is also possible to simplify co-ordinate dimensions by using a table to identify features and positions.

Figure 4-18: Coordinate dimensioning

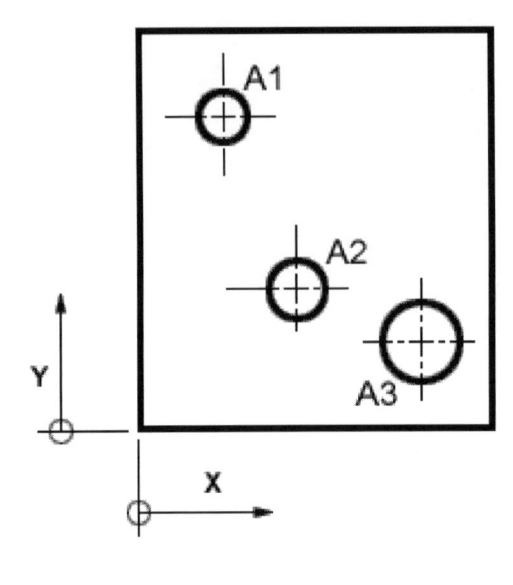

HOLE	X	Y	⌀
A1	100	25	25
A2	50	40	15
A3	100	20	15

7. *Location dimension*

Holes can be located by dimensioning the center of circle and with a note. They can also be located with coordinate dimensions from the center lines. Equally they can be located with angular dimensions and the radius from a pole point.

Center dimension Coordinate dimension Angular dimension

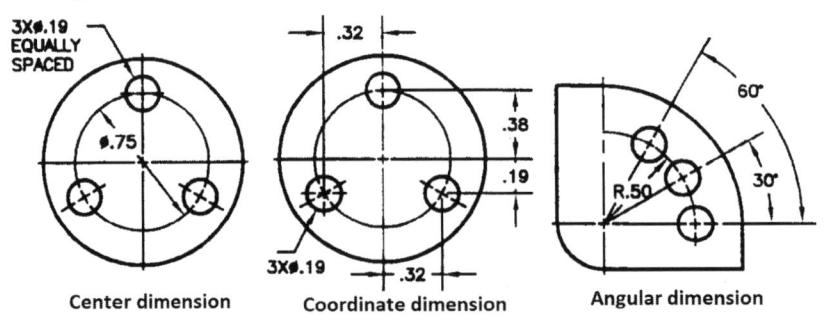

Figure 4-19: Location dimensions

8. *Dimensioning small features*

When dimensioning small features, placing the dimension arrow between projection lines may create a drawing which is difficult to read. In order to clarify dimensions on small features any of the above methods can be used.

→|5|←   →|⁵|←   →|⁵|←   →| |←5

Figure 4-20: Coordinate dimensioning

9. *Dimensioning small circles*

Method 1: Figure 4-21 below shows two common methods of dimensioning a circle. One method dimensions the circle between two lines projected from two diametrically opposite points. The second method dimensions the circle internally.

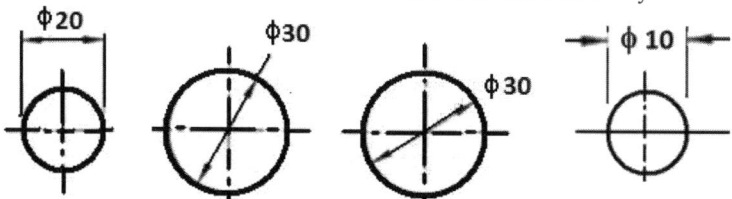

Figure 4-21: Small circle dimensioning

Method 2: When the circle is too small for the dimension to be easily read if it was placed inside the circle the figure above is used. A leader line is used to display the dimension.

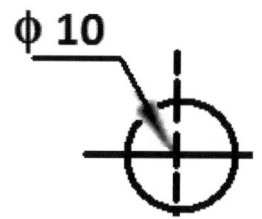

Figure 4-22: Small circle dimensioning

Method 3: The final method is used to dimension the circle from outside using an arrow which points directly towards the center of the circle. The first method using projection lines is the least used method. But the choice is up to you as to which you use.

Dimensioning holes: When dimensioning holes the method of manufacture is not specified unless they are necessary for the function of the product. The word hole doesn't have to be added unless it is considered necessary.

4 HOLES Ø7 x 10 DEEP

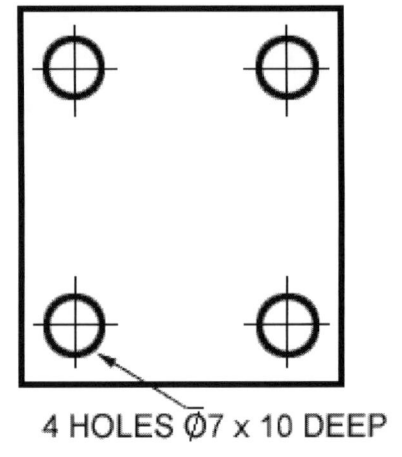

Figure 4-23: Small circle dimensioning

The depth of the hole is usually indicated if it is not indicated on another view as indicated in the figure above.

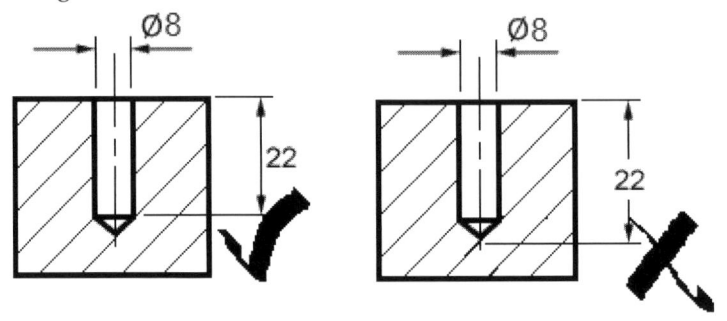

Figure 4-24: Hole dimensioning

The depth of the hole refers to the depth of the cylindrical portion of the hole and not the bit of the hole caused by the tip of the drip.

Radii dimensioning: All radial dimensions are preceded by capital letter R followed by a value. All dimension arrows and lines should be drawn perpendicular to the radius so that the line passes through the center of the arc. All dimensions should only have one arrowhead which should point to the line being dimensioned. There are two methods for dimensioning radii.

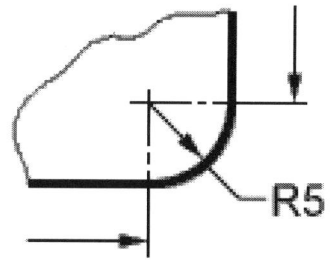

Figure 4-25: Radii dimensioning

Figure 4-26 above shows a radius dimensioned with the center of the radius located on the drawing.

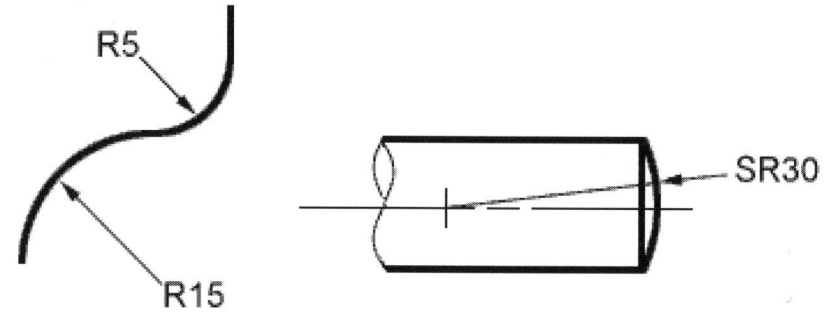

Figure 4-26: Radii dimensioning

Figure 4-26 above shows two ways of how to dimension radii which do not need their centers located.

Spherical dimensions: The radius of a spherical surface (e.g. the top of a drawing pin) when dimensioned should have an SR before the size to indicate the type of surface.

Dimensioning holes in drawings

In Figure 4-27 we have shown a hole that we have chosen to dimension on the left side of the object. The Ø stands for "diameter".

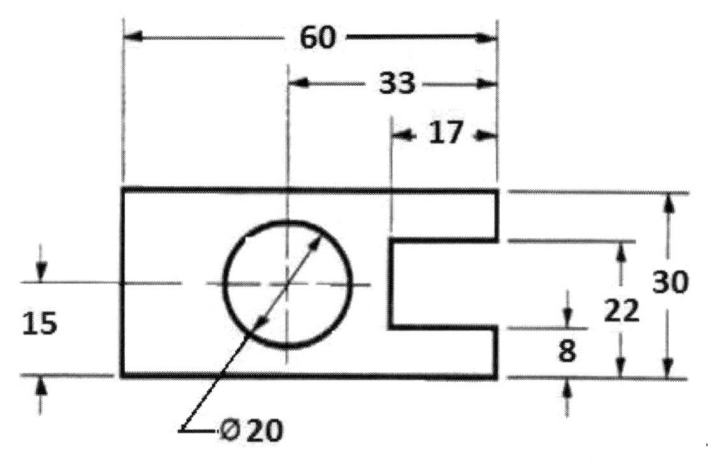

Figure 4-27: Example of a dimensioned hole

When the left side of the block is "radiuses" as in Figure 4-28, we break our rule that we should not duplicate dimensions. The total length is known because the radius of the curve on the left side is given. Then, for clarity, we add the overall length of 60 and we note that it is a reference (REF) dimension. This means that it is not really required.

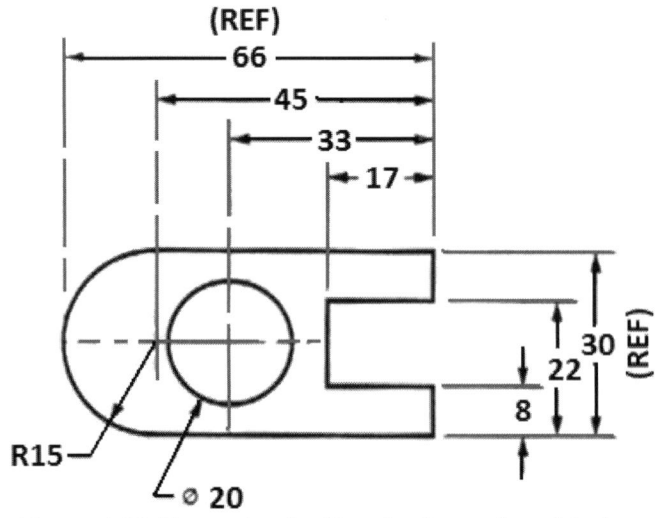

Figure 4-28: Example of a directly dimensioned hole

Somewhere on the paper, usually the lower right corner there should be placed information on the type of measuring system is being used (e.g. inches (imperial unit) or millimeters (metric)) and also the scale of the drawing.

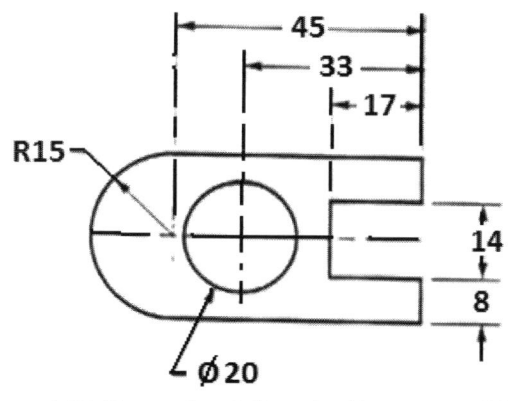

Figure 4-29: Example of directly dimensioned hole

This drawing is symmetric about the horizontal centerline. Centerlines (chain-dotted) are used for symmetric objects, and also for the center of circles and holes. We can dimension directly to the centerline, as in Figure 3-29. In some cases this method can be clearer than just dimensioning between surfaces.

Dimensioning cones

Two examples of dimensioned cones are shown in the Figure 4-29 below. The dimensions are placed on the triangular views

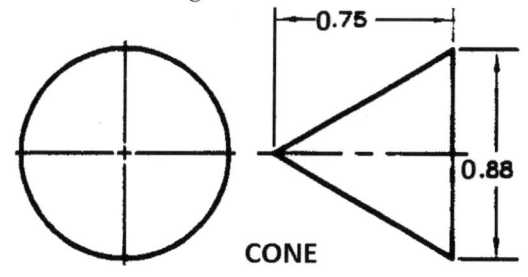

Figure 4-30: Cone dimensioning

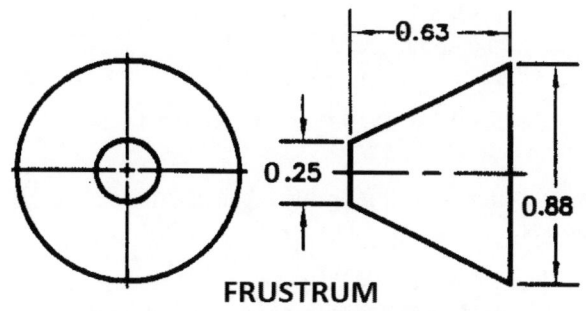

Figure 4-31: Frustrum dimensioning

Dimensioning isometric drawings

We have "dimensioned" the object in the isometric drawing in Figure 4-32. As a general guideline to dimensioning, try to think that you would make an object and dimension it in the most useful way. Put in exactly as many dimensions as are necessary for the craftsperson to make it -no more, no less. Do not put in redundant dimensions.

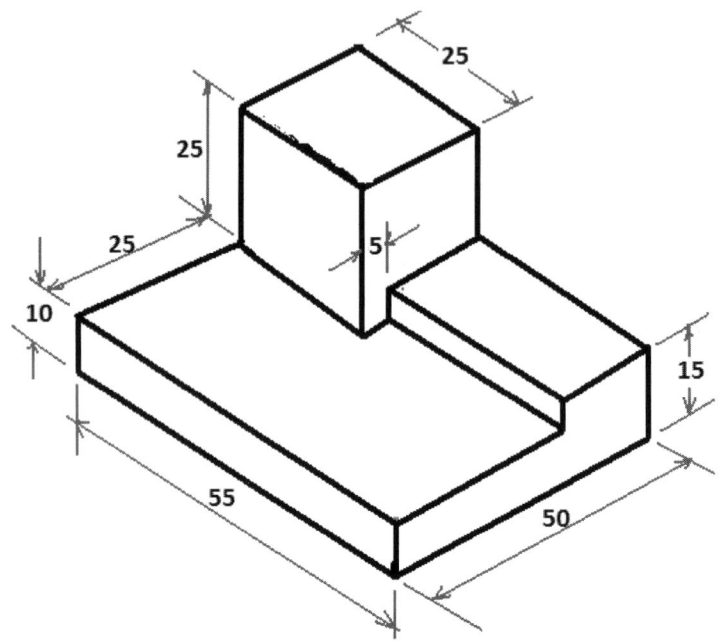

Figure 4-32: An isometric block

4.8 Dimensioning best practices

The followings are practices of dimensioning that meets the standards of drawing practice.

1. All letters in dimension must be All CAPS!
2. All figures must be in decimals
3. Select a front view that best describes the part
4. Remove hidden lines unless absolutely necessary to describe the shape of the object
5. Consider datum and dimensioning scheme based on
   a. Feature relationship
   b. Manufacturability and inspection
   c. Reduce math for machinist
6. Do not duplicate dimensions, use reference dims if necessary to duplicate

7. Do not dimension to hidden lines
8. Place dims between views if possible
9. No dims on body of part. Offset .38"inch from object outline
10. Place all dims for same feature in one view if possible
11. Dim lines cannot cross dim lines
12. Dim lines should not cross extension lines
13. Extension lines can cross extension lines
14. Use center marks in view(s) only where feature is dimensioned
15. Use centerlines and center marks in views only if feature is being dimensioned or referenced otherwise omit.
16. When multiples of the same features exists in a view, dimension only one of the features and label the dim as "Number X"DIM meaning that the feature exists in that view "Number" times. For example, "4X 0.250"implies that in the view, there exists 4 like dimensions for the dimensioned feature
17. Minimize use of centerlines between holes etc, they add little value and clutter the object being drawn.

Exercise

1. When do you place numbers and arrows outside the extension lines?
2. Sketch a drawing indicating all properties of dimensioning
3. List five types of dimensioning and indicated them with sketches
4. Sketch a drawing indicating all properties of dimensioning
5. List five types of dimensioning and indicated them with sketches
6. What does *unidirectional* mean when talking about dimension text? When is this method used?
7. What does *aligned* mean when talking about dimension text? When is this method used? Show two methods of dimensioning angles?
8. Draw and completely dimension the part given in two views below

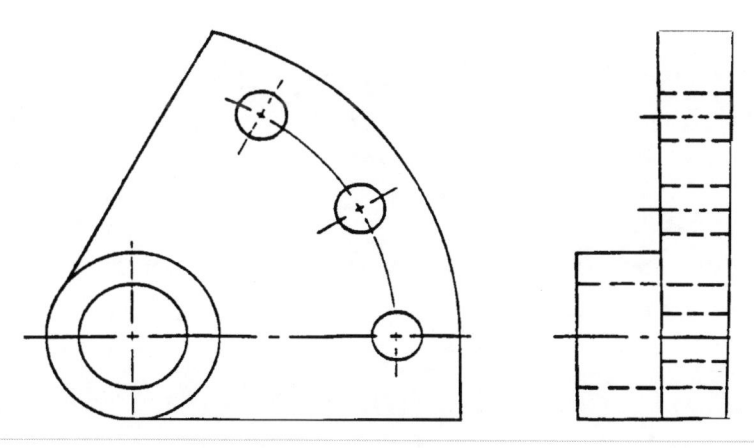

9. Reproduce the views represented below and completely dimension the part

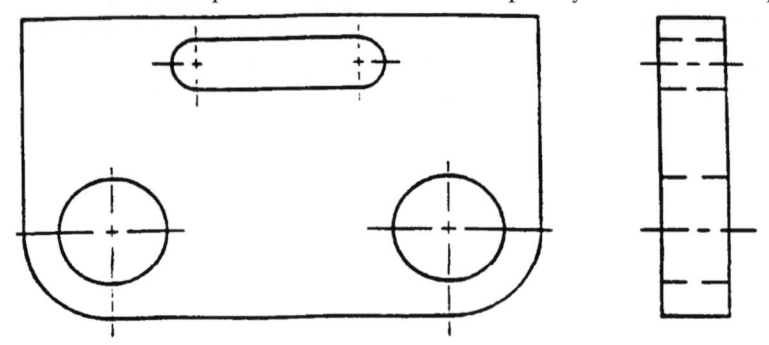

10. Are these two drawings the same? Which one would you rather detail

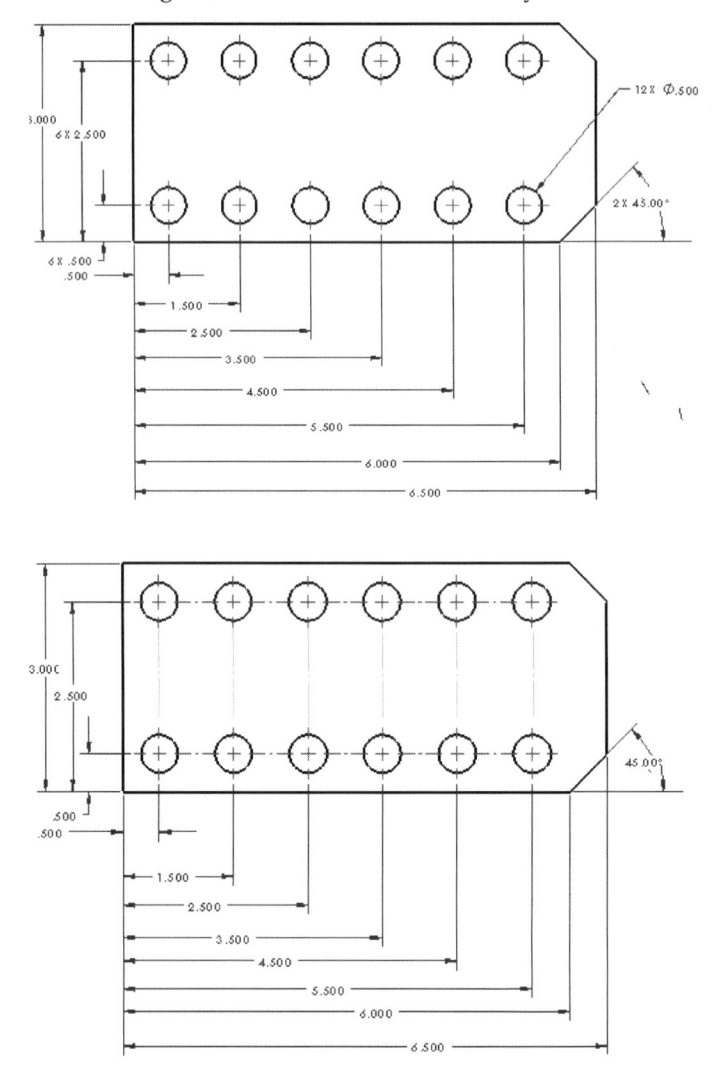

SECTION 2
Drawing Geometry & Projections

Introduction

Engineering/technical drawing is simply a means of communicating shape, size, position and proportion, features and precision of physical objects. One of the best ways to communicate one's ideas is through pictures, graphic illustration or drawings. Details of engineering innovations and technical inventions are hid in drawing for the purpose of safeguarding them. The following descriptions can also be used to describe the field:

- It is a universal language of engineering used in design processes for solving problems quickly and accurately by visualizing objects and conducting analysis.
- It can also be said to be a graphic representation of objects and structures used to solve problems which involve special relationships.
- It is a mode of thinking in which two-dimensional projections are used to visualize three-dimensional situations.
- It is also a means of describing and defining processes which verbal expressions cannot adequately conveyed.
- It is therefore an extension of language, and as such, an essential part of education in a technological society.

Drawings are used in all fields of Engineering (Mechanical, Civil, Architectural, Electrical, Aerospace, etc.) and agriculture. The types of drawings we shall be creating in this class are mechanical, but the concepts are all transferable to all other engineering fields. Engineering Drawing is concerned with imparting precise information about an object hence it is understandable that neatness and accuracy should play an important part in its practice. Beautifully copied engineering drawings are of little value if the principles behind the work are not fully understood and applied.

These principles can be learned and practiced by using freehand, mechanical aids, or computer aided design (CAD) methods. The basic concepts of engineering drawings practices have been introduced in *section one*, and meant for first semester teaching. *Section two* is the continuation of the basic principles guiding engineering drawing in second semester. The attempt here is to simplify the content to just what you need for classroom teaching.

At the completion of this section, the student should be able to tackle problems in;

1. *Projection of solids*

   a. Identify and define different types of protections - front vertical plane, horizontal plane, side vertical plane and auxiliary.
   b. Draw orthographic views of points, lines from pictorial views and vice-versa.
   c. Explain the difference between first and third angle projections
   d. Make composite drawings from orthographic views
   e. Differentiate between different pictorial views of objects
   f. Make pictorial drawing of objects

2. *Free hand sketching*

   a. Explain free-hand sketching. State the purpose and principles of free-hand sketching.
   b. Demonstrate the techniques of drawing free hand straight lines, circles, areas and angle.
   c. Make free hand sketches of pictorial and orthographic projections of simple objects, common tools and simple machine parts.
   d. Make sketches of simple three dimensional objects.

Chapter 5

Geometrical Figures: Properties and Construction

5.1 Introduction

Plane geometry describes the properties and construction features of elements of geometrical figures. Such geometrical elements includes: points, lines, surfaces and planes. In the basic engineering drawing we mainly discuss geometrical drawing as the art of representation of objects on a drawing sheet and it is the foundation of all engineering drawing. Figure 5-1 below shows examples of various types of plane figures.

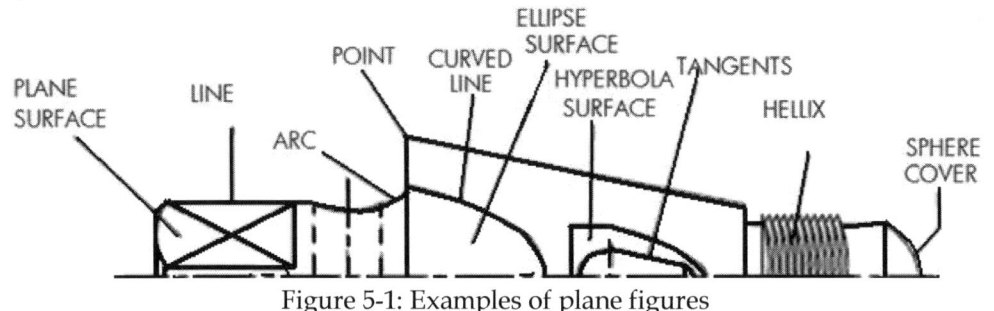

Figure 5-1: Examples of plane figures

5.2 Point

Point is a non-dimensional geometrical element. It has no area and magnitude and is often indicated by a dot (.) or plus (+).They occurred at interception of various lines indicating a position or reference of a figure.

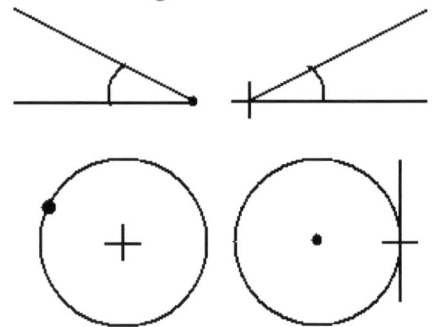

Figure 5-2: Examples of point or reference

5.3 Lines, properties, application and construction

A line is a 1-dimensional geometrical element generated by the movement of a point in various directions. A Line is equally a locus of points equidistant from a common reference path traced out by the line. A Line is described by its length and thickness however, no area.

Lines in an engineering drawing signify more than just the geometry of the object but a description of relative position and state of such part so represented and it is important that you use the appropriate line types. Generally; technical drawing is the expression of bodies (or objects) by lines. Material work-pieces are composed of variable geometric components. Sides and surfaces of these components are visible but some of them cannot be seen because they are behind the back sides.

Figure 5-3 below illustrates different types of lines, drawn in various directions, and a combination of these lines generates other geometrical elements.

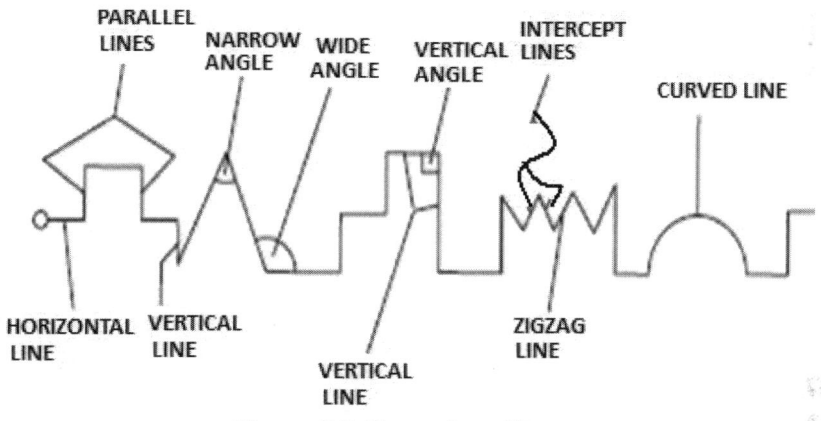

Figure 5-3: Examples of lines

To obtain full and precise information about the piece, a drawing should be done by using variable lines instead of using one particular line style. Moreover, these lines should be drawn at same thickness and shape by everyone.

Line thickness

For most engineering drawings you will require two line thicknesses, a thick and thin line. The general recommendations are that thick lines are twice as thick as thin lines.

A thick continuous line is used for visible edges and outlines.

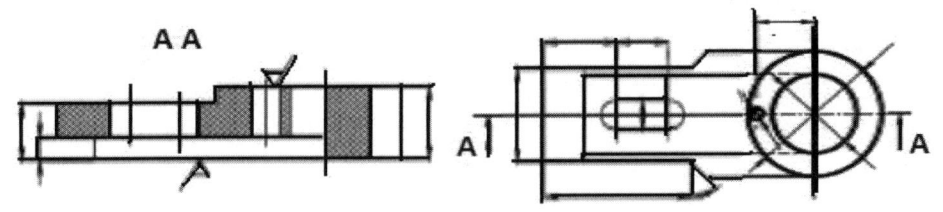

A thin line is used for hatching, leader lines, short center lines, dimensions and projections.

Figure 5-4: Drawing lines

Considering the variability of lines in a particular drawing, consider the Figure 5-4 above. Same line shapes and thicknesses are used in this drawing. Therefore it is very difficult to have an idea of the shapes and dimensions of the piece.

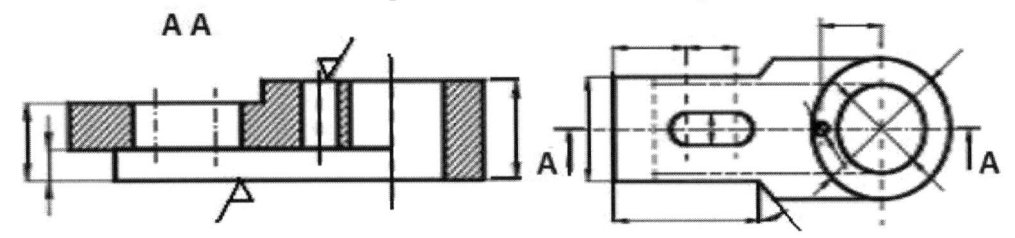

Figure 5-5: Drawing lines differing in shapes and dimension

The Figure 5-5 above shows the same piece, which is drawn by using variable lines differing in shapes and thickness to show visible and invisible lines, axis and dimensions. In this way, one has a full idea of the piece.

5.3.1 Line styles

Other line styles used to clarify important features on drawings are:

Thin chain lines are a common feature on engineering drawings used to indicate center lines. Center lines are used to identify the center of a circle, cylindrical features, or a line of symmetry. Center lines will be covered in a little bit more detail later in this tutorial.

Dashed lines are used to show important hidden detail for example wall thickness and holes.

5.3.2 Forms of lines

Straight line is the shortest distance between two points. It has the same direction throughout its length and has no breadth or thickness.

A *Curved line* changes its direction throughout its length.

Parallel lines are the same distance apart and cannot meet however far they are produced.

Horizontal lines are lines produced parallel to or along the x-axis while the vertical lines are parallel to or drawn along the y-axis.

Oblique/Incline lines are neither parallel to the x- or y- axis. They are drawn at an angle to the principal planes.

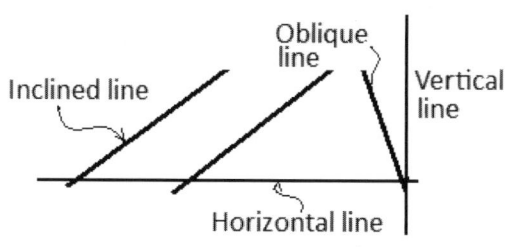

5.3.3 Types of lines

According to the "TS 88 technical drawing standard" published in 1978, lines are classified into the following nine types and are recommended for use in construction:

1. *Construction line*: This line is a faint and narrow or thin line used in preliminary construction details of the intended drawing development. It is represented with a thin, faint, continuous line as shown

2. *Finished detail lines*: These are bold lines used in final detail drawing of the sketchy drawing produced with construction detail lines. They are also continuous

3. *Dimension lines:* Drawing specifications including size, shape, etc. are represented with dimension lines. They are double headed arrow lines bounded by short vertical lines at the ends. The lines have definite lengths

4. *Center lines:* These are lines of division indicating the symmetrical nature of an object. Center lines depict the center of any cylindrical-shaped object whether it is a cylinder or hole. They are shown as a long line followed by a short line, followed by a long line.

5. *Projection lines:* These are used to project images or objects onto a plane. They are represented by long dashes with one or two short dashes or dots in-between them.

6. *Hidden detailed lines:* Objects that are not visible to the naked eye are represented by short continuous dashes. Hidden lines depict invisible edges inside an object. The edges you would not see looking at the object with your naked eyes.

7. *Broken line*: This is a continuous line cut by an inverted letter 'Z'. This indicates a cut in the original drawing when an enlarged detail of a section of a drawing is required. Other pars are cut off by the broken line.

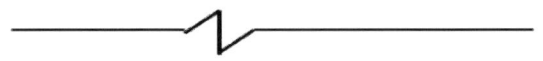

8. *Section line:* This line indicates a cut through an object thereby revealing inner details. Sectional views use both thick and thin line with the hatching carrying on to the very edges of the object.

    Two illustrations (Figures 5-6 & 5-7) below demonstrate two methods of drawing a threaded section. Note the conventions in each of the drawings. In Figure 4-6 the inner thread is drawn with continuous bold line but the section is hatched. In Figure 5-7, the hidden detail is drawn as a thin dashed line.

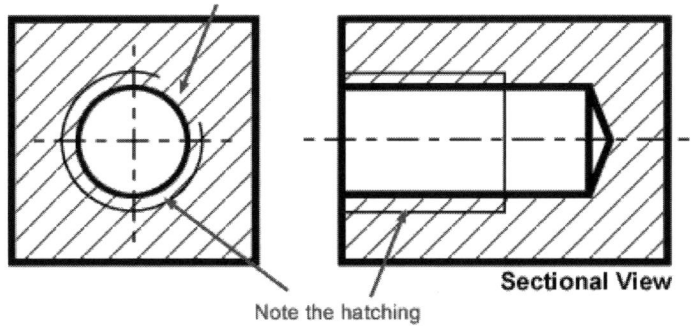

Figure 5-6: Hatching treaded sections

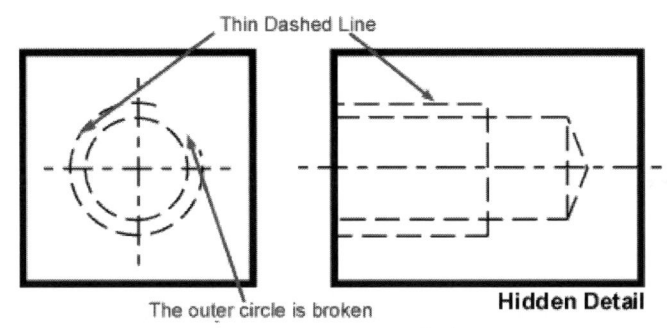

Figure 5-7: Drawing treaded sections

5.3.4 Line applications

Areas of applications of the different types of line are listed in the following table:

	Type of lines	Areas of application
A	Continuous Line (Thick)	A1- Surroundings and sides of the matters. A2- End of the A Screw
B	Continuous Line (Thin)	B1- Backside section line B2- Measure lines, guide lines B3- Simplified axis lines B4- At diagonal lines which are used to state plane surface B5- to State the Code of the Places

C1 C2	Free hand lines Zigzag Line (Thin)	C1-To state the place that limits section and appearance of matter or to state the place tore off. C2- It is used when free hand lines are drawn by tool.
D	Dash Line (Thin)*	D1- Invisible surrounding and sides of the matter.
E	Dash Line with Point (Thin)	E1- Axis lines of symmetrical drawing E2- In Front of Section Planes
F	Section plane with thick ends and thin mid points.	F1- To draw the traces at section plane
G	Dash Line with Point (Thick)	G- To state the place which will processed additionally. (To coat, to harden, etc.)
H	Dash line with two points (thin)	H1- To show the surroundings of neighbor pieces H2- To state the secondary situation of moving pieces. To state the center of gravity

5.3.5 Line drawing

Thickness of lines should be drawn according to standards as stated below.

a. *Continuous and thick lines* should be drawn with a B or 2B pencil.
b. *Continuous and thin lines* should be drawn with an H or 2H pencil.
c. *Dash lines* should be drawn at equal spaces and thickness. They should be 3~6 mm, or 0.8~1.5 mm

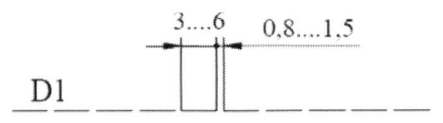

Figure 5-8: Drawing dash line

d. *Dash lines with point* should be drawn according to the size of the picture with mentioned sizes below.

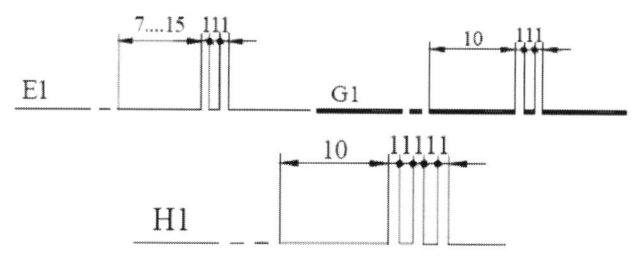

Figure 5-9: Drawing dash line with size

e. *Intersected continuous lines* should not be overflowed or uncompleted at the intercept points. Thicknesses should be same and corners should be sharp.

Figure 5-10: Drawing intersected continuous line

f. *Junctions of circle arcs and lines* should be tangent.

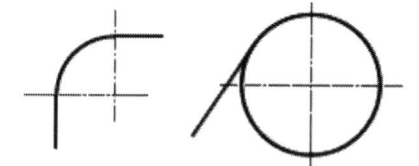

Figure 5-11: Drawing tangent to curves

g. Minimum space between two parallel lines should not be less than two times of the thick lines.

d= Thickness of lines

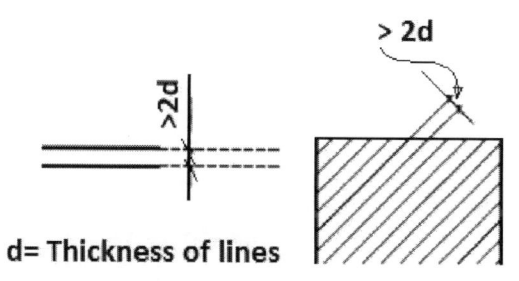

Figure 5-12: Drawing line spacing

5.3.6 Line construction

1. *Drawing parallel line with compass*

Drawing parallel lines can be achieved in two ways:

a. Drawing parallel line to a line from any P point (outside from the line)
b. Drawing a parallel line to a line with a known distance, "a"

Each of these methods is examined as follows;

a. *Drawing parallel line to a line from any P point (outside from the line)*

Method 1

i. Accept point p as center. Open the compass with arc r, intercept *ab* line and obtain point c.
ii. Accept point c as center, don't move the compass. Draw another arc that cross point p and intercept *ab* line, obtain point d.

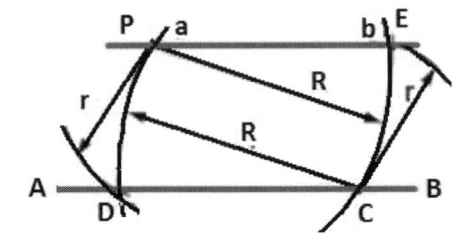

Figure 5-13: Drawing parallel line with compass

iii. Open the compass as *PD* arc. Put the compass to point C and intercept arc b, find point e.
iv. Connect the point P with point E.

Method 2

i. Draw any line crossing point p, intercepting line *ab*.
ii. Accept point c as center. Open the compass as *cp*, draw an arc, and obtain point d.

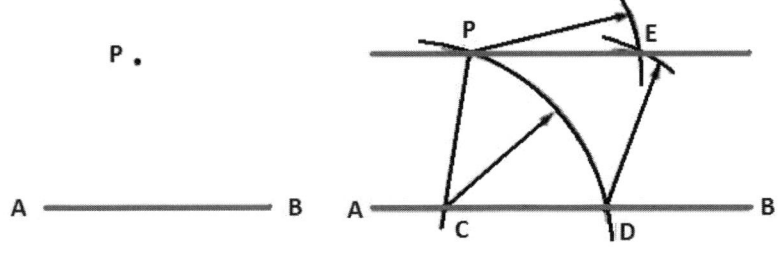

Figure 5-14: Drawing parallel line with compass

iii. Accept point *p* and *d* as center, respectively and find point *e* with intercepted arcs.
iv. Connect point *p* and *e* as the required line.

b. *Drawing a parallel line to a line with a known distance, "a"*

i. Open the compass as "*a*".
ii. Mark any *c* and *d* points on *ab* line.

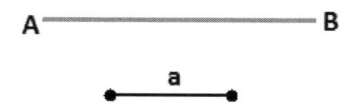

A ——————————— B

a

Figure 5-15: Drawing parallel line with known distance

iii. Draw two arcs by accepting *c* and *d* points as center, respectively.
iv. Draw tangent *EF* to these arcs as the required line.

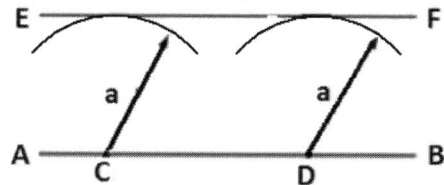

Figure 5-16: Final drawn parallel line

Drawing of vertical lines

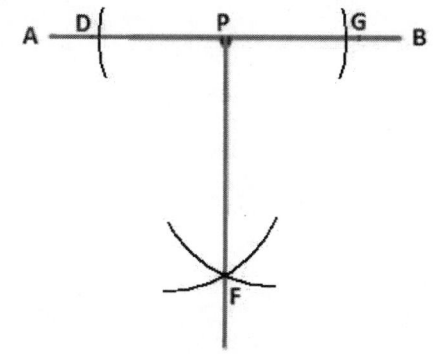

a. Method 1: Drawing vertical lines with compass

To draw vertical line from the point on a line

i. Accept *p* point as center. Draw *d* and *g* point on line *ab*.
ii. By accepting *d* and *g* points as center, respectively, draw two arcs that are intercepted outside from *ab* line and form *f*
iii. Connect point *d* and *f*.

Figure 5-17: Drawing vertical line

b. *Method 2: Drawing a vertical line at the end of a line*

Method 1:

i. Accept point P as center (Figure 5-18).
ii. Draw arc *r* and mark point *b*.
iii. Don't move compass angle, accept point *b* as center, and draw an arc crossing point *p* and previous arc. Obtain point *c*.

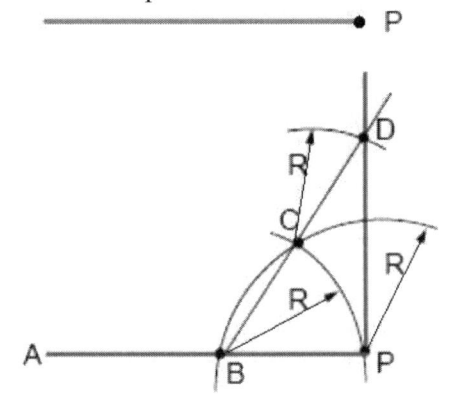

Figure 5-18: Final drawn parallel line

iv. Connect point *b* and *c* and prolong this new line.
v. Accept point *c* as center; draw an arc intercepting *bc* line.
vi. Mark point *d* at the intersection point.
vii. Connect point *p* and *d* as the required vertical line.

Method 2:

i. Accept point *p* as center. Open the compass as *r* amount landmark point *b*.
ii. Don't move the compass; accepting *b*, *c* and *d* as center, respectively; draw the arcs intersecting each other. Obtain point *e*.
iii. Connect point *p* and *e*.

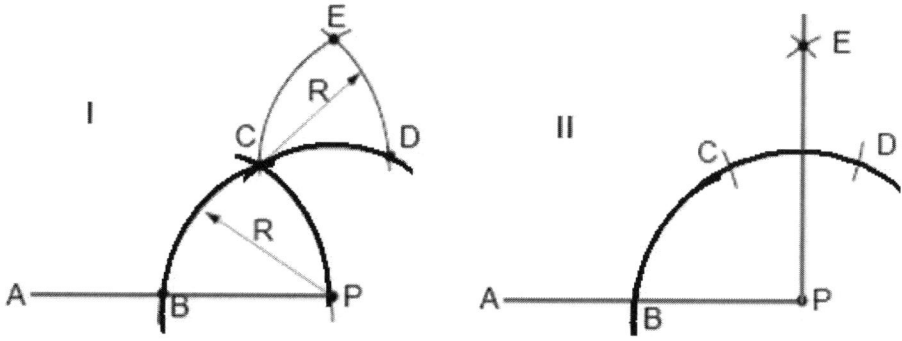

Figure 5-19: Final drawn parallel line

c. Method 3: Drawing a vertical line to a line from an outside location

i. Accept point p as center. Draw an arc crossing line AB. Mark points C and D.
ii. Accept point C and D as center, respectively. Draw two arcs intercepting each other. Mark point E.
iii. Connect point E and P.

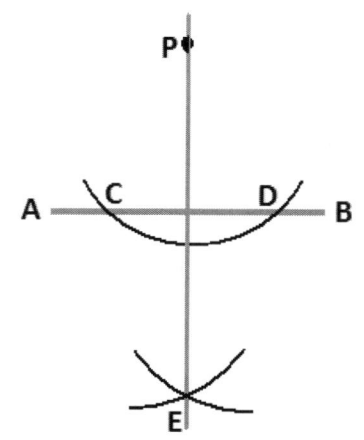

Figure 5-20: Drawing line from outside location

Dividing line to parts (segments)

a. Dividing a line to two, four and eight equal parts:

   i. Open the compass as little more than half of the line.
   ii. Accept point A and B as center, respectively. Draw intersecting two arcs.
   iii. Connect the intercepting point and obtain point c. In this way, you can divide the line AB to two equal pieces

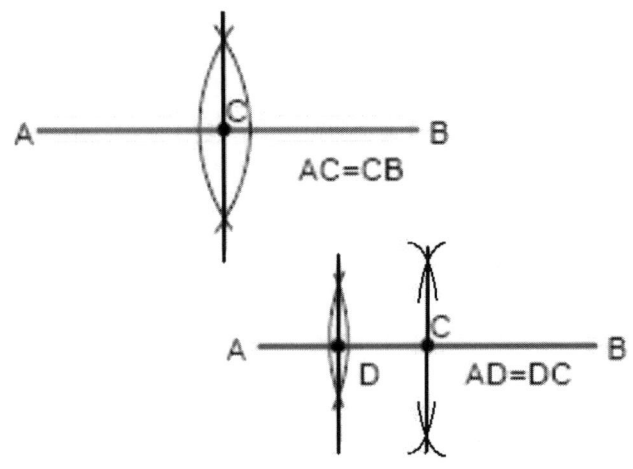

Figure 5-21: Line division process

iv. Repeat the same procedure for *ac* line. Obtain point *d*.
v. Repeat the same procedure for *ad* line, obtain point *e*.

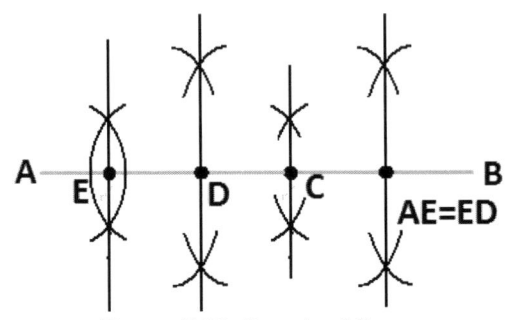

Figure 5-22: Required lines

b. *Dividing a line into specific number of segments:*

Suppose line segment AB is given and that it is required to divide AB into seven equal parts.

Procedure:

i. Through point A draw line m and then lay off seven equal distances, starting at point A.
ii. Join points E and B and then draw parallels to EB through the points on m.
iii. The points of intersection of these parallels with line segment AB divide it into the required seven equal parts.

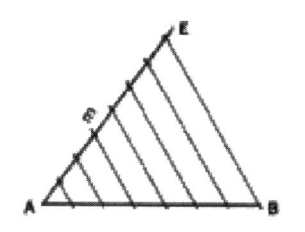

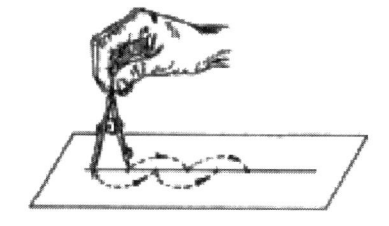

Figure 5-23: Line division using compass

In laying off distances on a line segment, use needle-point dividers, alternating the rotation of the dividers as shown.

c. *To divide a line segment into a given ratio*

Procedure:

i. Let AB represent the given line segment, and let the given ratio be 4:5:7.
ii. Draw line m through point A and at any convenient angle with AB.
iii. On m lay off distances AC = 4 units, CD = 5 units, DE = 7 units.
iv. Draw BE and then draw lines through C and D parallel to BE, cutting AB in points C' and D', which determine the required segments of AB

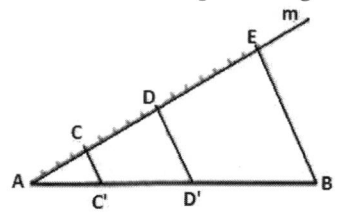

Figure 5-24: Line division into given ratio

d. *Constructing mean proportional (geometric mean) line to given line segments*

To construct a line segment which is the mean proportional (geometric mean) to two given line segments you can precede as follows

Procedure:

i. Suppose the given segments are m and n.
ii. On line AB lay off consecutive segments equal to lengths m and n.
iii. Construct a semicircle on the total length (m + n) as a diameter.
iv. At the common point of the segments construct a perpendicular to the diameter.
v. Line E is the required mean proportional.

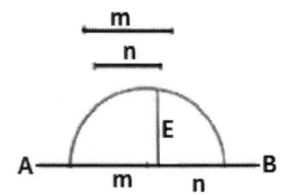

Figure 5-25: Line division into given ratio

The student should prove this by showing that g squared = m X n.

e. *To divide a straight line in extreme and mean proportion*

Procedure:

i. Assume the given line is AB.
ii. At point B lay off line BC at 90° to AB and equal to AB/2.
iii. With C as center and radius CB, draw an arc cutting line AC at point D.
iv. With A as center and radius AD, draw an arc cutting line AB at point E, which divides line AB in extreme and mean proportion (i.e., the square on segment AE is equal to the rectangle having sides AB and EB).

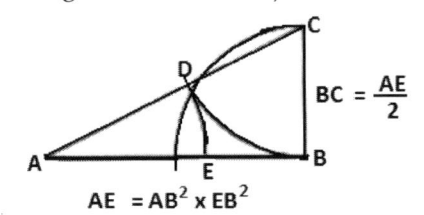

Figure 5-26: Line division in proportions

5.3.7 Loci: Properties and construction

When a point moves according to a given law its path is said to be a locus. Also a point moving such that it is always at a constant distance from a second point traces out a circular part as its locus. The full set of points gives what is referred to as a path.. A locus is a set of points all of which share some common property, for instance, the locus of point which moves so that its distance from a fixed straight line is constant is a straight line parallel to the fixed line.

The word 'loci' is the plural form of the word locus, which is a Latin word meaning place. Hence, the word, loci, is used in mathematics to mean a collection of points that meet some given condition or conditions.

A locus may be a point, a line, a curve or a region. The important point is that all the points that make up the locus have to satisfy the same rule or condition. For example, you might be asked to draw the locus of points that are a certain distance from a given point or line.

Examples of loci

Many important geometrical figures and curves may be regarded as loci. Example includes ellipse, parabola, hyperbola, cycloids etc. The properties and construction of some of these figures will be discussed under conic sections.

The following special cases of description of locus exist in drawing:

Locus of all points equidistant from a given point

Equidistance means points of equal distance. This could be applied in "real life" to the situation where a goat is tie to a pole and allowed to tether say 2m from the pole. If we keep on marking such points we would get a circle around the pole, the radius of which was 2m. The dashed line in figure below is the locus we want.

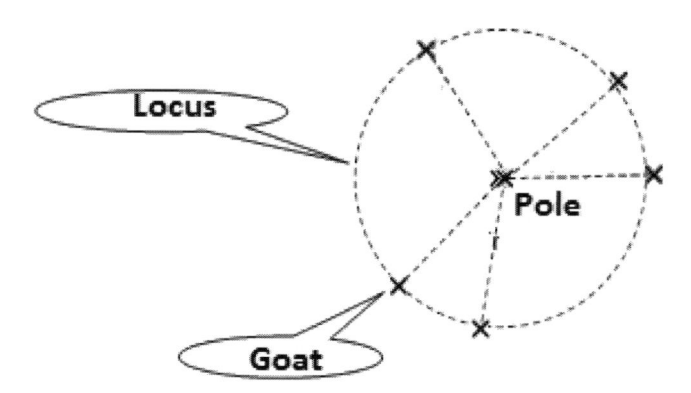

Figure 5-27: Locus of point equidistant from given point

Locus of all points equidistant from a given line segment

In this situation, suppose we have a wall and we are told an object is hidden 5m from the wall. We would move perpendicular 5m from the wall and not at an angle to it. When a plot of the collection of points is plotted, it looks like the shape in figure below.

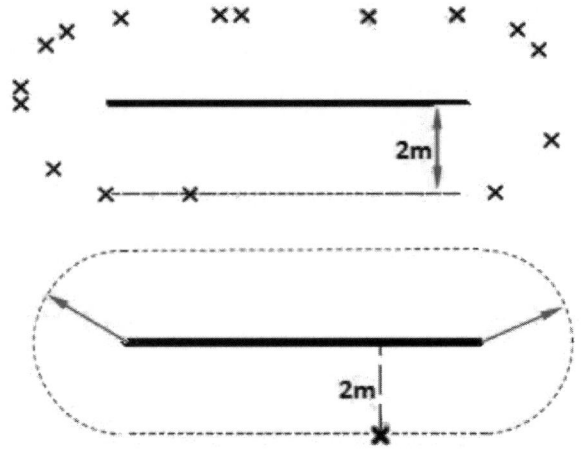

Figure 5-28: Locus of points from a given segment

Locus of points equidistant from two points

If you are required to draw the locus of the points that are equidistant from A and B, all the points must be the same distance from A as from B. Hence, the locus is the perpendicular bisector of the line AB.

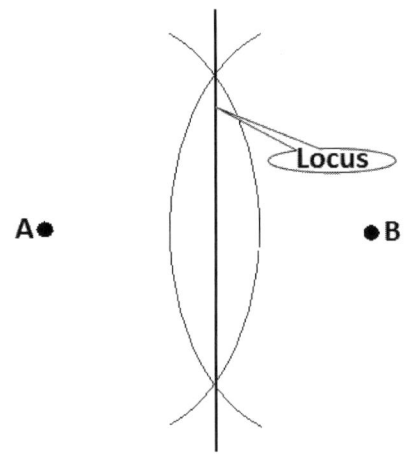

Figure 5-29: Locus of points from given points

Also consider drawing the locus of points that are 1cm from the circle shown below. The locus is made up of 2 parts. 1 part consists of the points that are 1 cm from the circle and inside it; the other is those points that are 1 cm from the circle and are outside it.

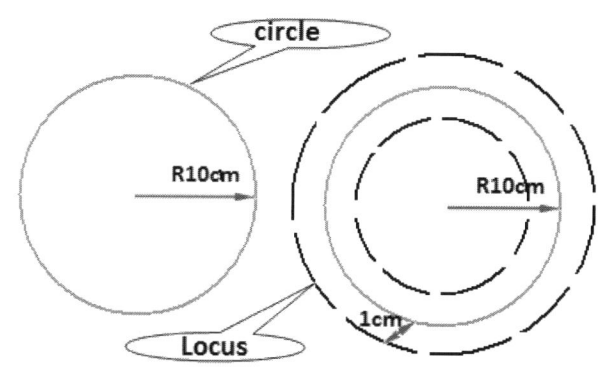

Figure 5-30: Locus of point round circle

Locus of all points equidistant from two intersecting lines

Here we have the bisector of the angle between the two lines –again a construction.

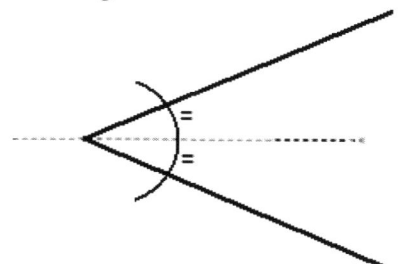

Figure 5-31: Locus of point from two interacting lines

5.4   Plane and solid loci

The characteristics demonstrated in lines as loci can also be studied for plane, spherical, conical, and cylindrical surfaces or by those of arbitrary conoids and spheroids, if we first establish the lemmas constituting each of these loci. These lemmas are explained below.

i. *Plane surface loci*: If a given surface is cut by as many planes as you wish, and the intersection of this surface with each of the planes is always a straight line, then the surface in question will be a *plane*.
ii. *Loci on spherical surfaces*: If a given surface is cut by an arbitrary number of planes, and the intersection of the surface with each of the planes is always a circle, then the surface in question will be a *sphere*.
iii. *Loci on spheroidal surfaces*: If a given surface is cut by an arbitrary number of planes, and the intersection of the surface with each of the intersecting planes is either a circle or an ellipse, but never another curve, then the surface in question will be a *spheroid*.

iv. *Loci on parabolic conoid or hyperbolic surfaces*: If a given surface is cut by an arbitrary number of planes, and the common intersections be either circles, ellipses, parabolas or hyperbolas, but never another curve, then the surface in question will be a *parabolic conoid* or *hyperbolic*.
v. *Loci on conical surfaces*: If a given surface is cut by an arbitrary number of planes and the common intersection be always a straight line, a circle, an ellipse, a parabola or a hyperbola, but never another curve, then the surface in question will be a *cone*.
vi. *Loci on cylindrical surfaces*: If a given surface is cut by an arbitrary number of planes, and the common intersection be either a straight line, a circle or an ellipse, but never anything else, then the surface in question will be a *cylinder*.

But loci often present themselves whose sections (cuts) are straight lines, parabolas and hyperbolas, and nothing else, as the analysis of the question will soon show. Therefore it is suitable, or perhaps even absolutely necessary for this study, to constitute a new species of cylinders having as parallel bases either parabolas or hyperbolas, and straight lines for their edges, self-parallel and joining the bases, by analogy with ordinary cylinders.

5.5 Plane figures

A plane is formed by at least three points or connection of one point and one line. A plane is always 2-dimentional. When the number of element forming a plane increases, shape and name of the plane will change.

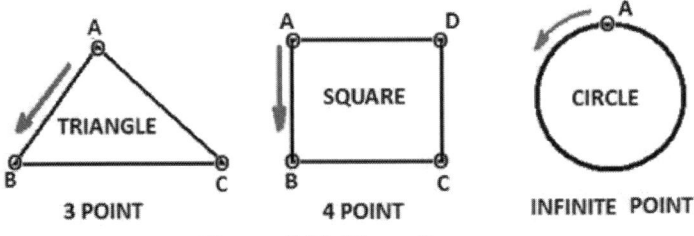

Figure 5-32: Plane figures

The connection of three points at certain conditions forms a triangle.
The connection of four points at certain conditions forms a square.
The connection of infinite points at certain conditions forms a circle.

Note: A connection is formed at the intersection of two lines (points A, B, or C) as shown in the Figure above. The description and properties of these plane figures are described as follows:

5.5.1 Angles and their properties

If two lines are pivoted as shown in the diagram, as one line opens they form an angle. If the rotation is continued the line will cover a full circle. The unit for measuring an angle is a degree, which is 1/360th part of the whole circle. The point where the two lines meet is called the pivot.

Types of angle

According to the situation of lines crossing each other, three types of angles occur. These angles are seen in Figure below.

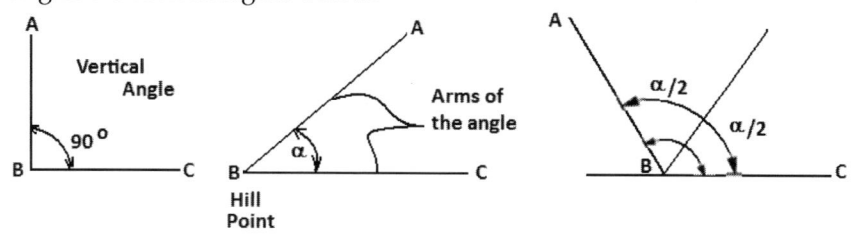

Figure 5-29: Types of angles

1. *Acute angle*: This is an angle less than 90°

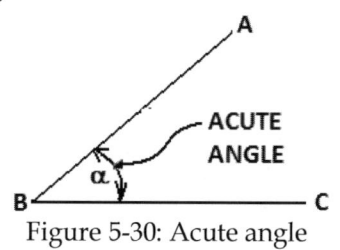

Figure 5-30: Acute angle

2. *Obtuse angle*: An angle greater than 90° but less than 180°

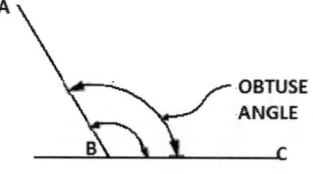

Figure 5-31: Obtuse angle

3. *Right angle* is an angle that is equal to 90°

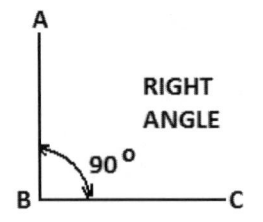

Figure 5-32: Right angle

4. *Reflex angle* is greater than 180° but less than 360°

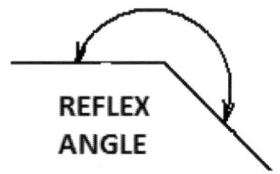

Figure 5-33: Reflex angle

5. *Complimentary angle*: When two angles added together make 90°. They are said to be Complimentary

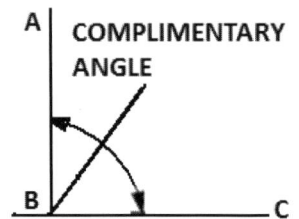

Figure 5-34: Complimentary angle

6. *Supplementary angle*: When two angles added together make 180 they are said to be supplementary

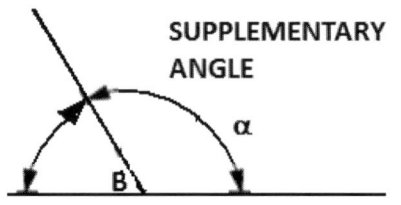

Figure 5-35: Supplementary angle

7. *Compass angles*: All angles that are based on marking out an angle of 60 or 90 degrees and then bisecting for smaller angles are called compass angles. Examples include 60, 90, 45 30, 75 15, 7.5 etc.
8. *Construction angles* are those that marked based on construction. Examples include all other angles besides those of compass angles mentioned above.

5.5.2 Triangles, their properties and construction

A triangle is a plane figure bounded by three straight lines. They are named or classified according to the length of their sides or the magnitude of their angles. In every triangle the sum of the three angles equals 180°

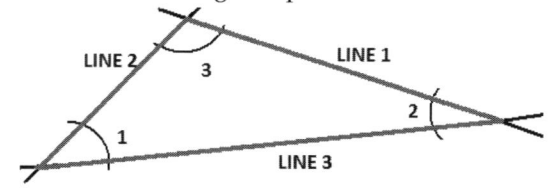

Figure 5-36: Triangle

The following types of triangles are common

1. *Equilateral triangle*; this has all its three sides and angles equal to each other. Each of the angles equal 60°. A line drawn from the one angle bisects the base line and at 90°. A perpendicular line drawn from the vertex to the base of the triangle is named the altitude.

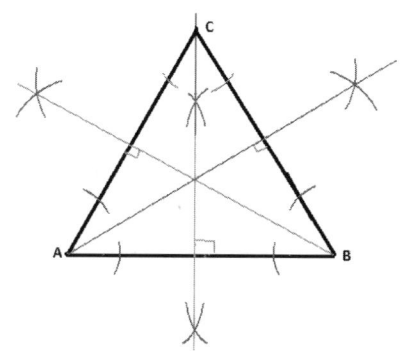

Figure 5-37: Equilateral triangle

2. *Isosceles triangle* has two of its three sides and angles equal.

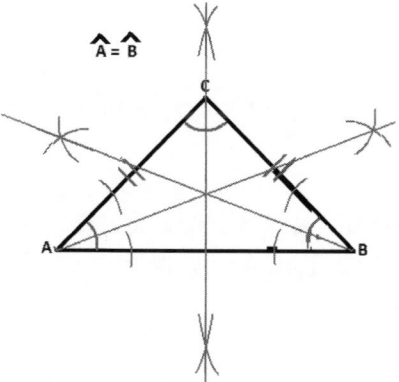

Figure 5-38: Isosceles triangle

3. *Scalene triangle* has all its three sides and angles not equal to each other. The triangle may be obtuse angled i.e. one of its angles is greater than a right angle (90°). The height or altitude is the perpendicular distance from the vertex. The bisectors of the angles are referred to as medians and they meet at a point *o* in the triangle.

An acute scalene triangle has each of its three angles less than a right angle (90°)

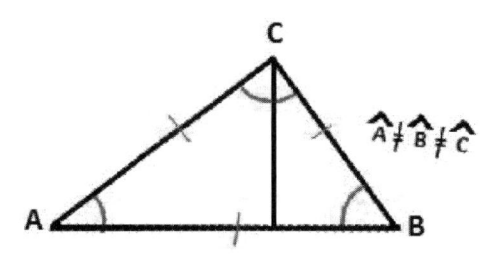

Figure 5-39: Scalene triangle

4. *Right angled triangle* has one of its three sides equal to 90°. The side facing largest angle (90°) is called the hypotenuse. The two remaining angles of a Right angled triangle are complimentary.

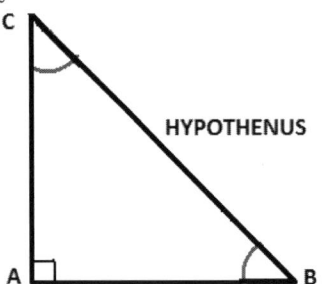

Figure 5-40: Right angled triangle

Construction of triangles

Construction of equilateral triangle

Method 1: Drawing of equilateral triangle-one side given. (With compass) proceed as follows:

1. Side *AB* is known. Open the compass as *ab*. Draw two arcs by considering *a* and *b* as center, respectively. Obtain point *c*.
2. Connect point *c* with A and B.

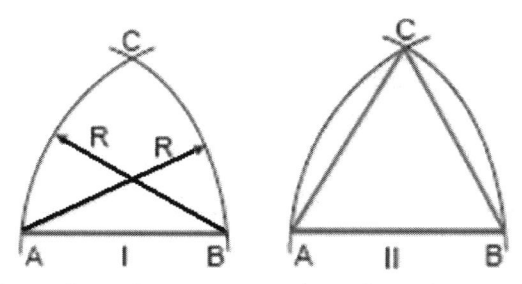

Figure 5-41: Construction of equilateral triangle

Method 2: Drawing of equilateral triangle in a circle (or, dividing a circle to three equal pieces) proceeds as follows:

1. Open the compass as radius (*r*) of the circle.
2. Accept the intersecting point of the circle with lateral or vertical axis of the circle as center. Draw an arc, crossing the circle at two points
3. The length between *a* and *b* points, obtained from previous step, is the beam length.
4. Point *c*, which is the opposite of the center, is connected with *a* and *b* points.

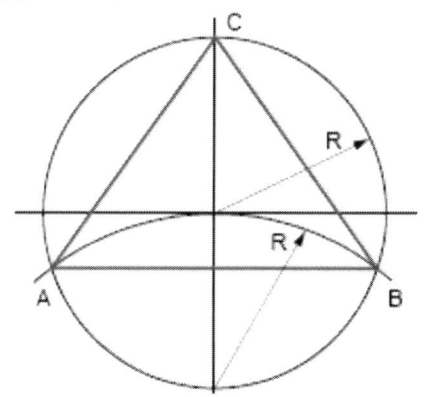

Figure 5-42: Construction of equilateral triangle

Construction of perpendicular (right angled) triangle

Method 1: Drawing of a perpendicular triangle of which two perpendicular sides are given:

1. Draw side *ab*.
2. Draw a vertical line at point *a*.
3. Mark *ac* side with the help of compass at this vertical line. Obtain point *c*.
4. Connect point *c* with point *b*.

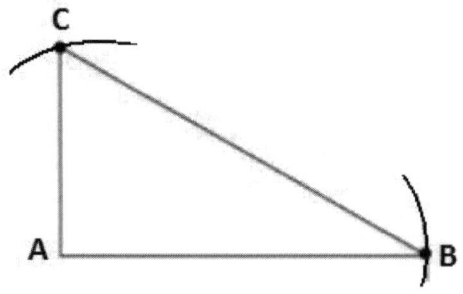

Figure 5-43: Construction of perpendicular triangle

Method 2: Construction of a triangle given the perimeter and the ratio of the sides

The procedure is as follows:

1. Let *AB* be equal in length to the perimeter
2. Divide *Ab* into the proportions say 3:4:5
3. With centers *c* and *d* and radii *ca* and *db*, draw the arcs to intersect in *e*
4. Join *ECD* the required triangle.

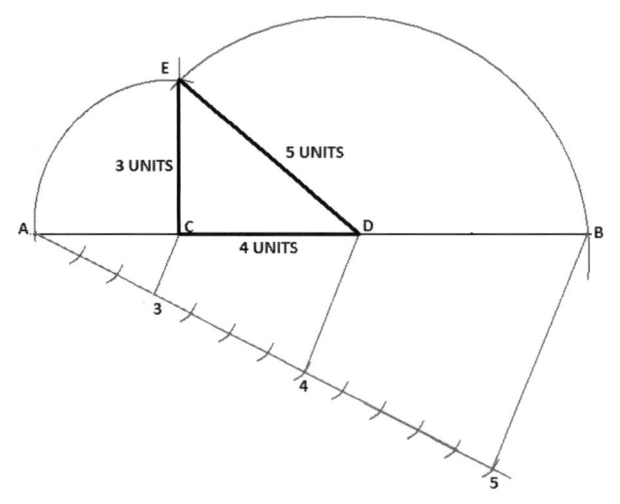

Figure 5-44: Division of triangle

Method 3: Drawing a perpendicular triangle in a circle.

The procedure is as follows:

1. Draw a circle with radius *r*.
2. Mark any point on the circle, as *a*.
3. Connect the point *a* with points *b* and *c*, which are the crossing of circle with radius *r*

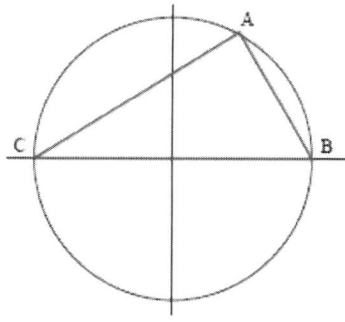

Figure 5-45: Construction of perpendicular triangle in a circle

Constructing isosceles triangle giving the perimeter and the altitude

Proceed as follows:

1. Draw AB be equal to half of the perimeter
2. Erect the altitude BC and join AC
3. Draw perpendicular bisection of ac to cut AB at d
4. Make BE equal to BD
5. CED is the required triangle.

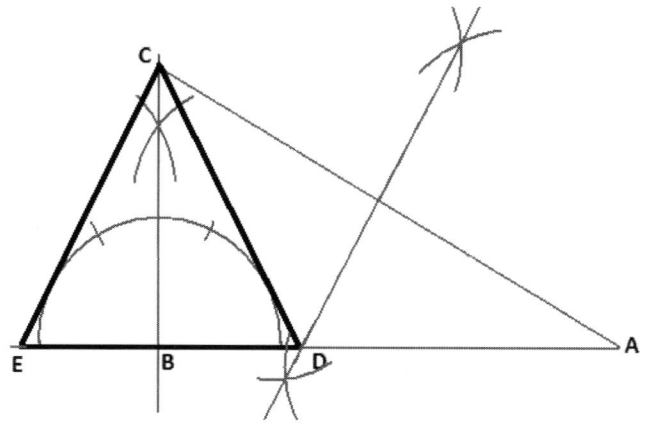

Figure 5-46: Construction of isosceles triangle giving the altitude

To construct a triangle with known lengths of the sides

Procedure:

1. Suppose m, n, and s are the given lengths.
2. With the end points of m as centers and radii equal to n and s, respectively, draw intersecting arcs, locating point A.

3. Join A with the end points of m to complete the construction of the triangle.

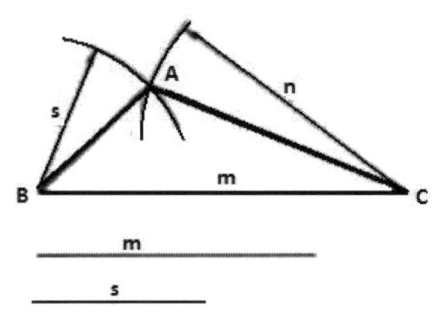

Figure 5-47: Constructing triangle with known sides

Is it possible to construct a triangle with 7, 3 1/2 and 3 inches as the lengths?

To construct a right triangle when the lengths of the hypotenuse and one side are known

Procedure

4. Let line m and n represent the given lengths.
5. Construct a semicircle with diameter AB equal to length m.
6. With A as center (could use B) and radius equal to length n, draw an arc cutting the semicircle at point C.
7. Triangle ABC is the required right triangle.

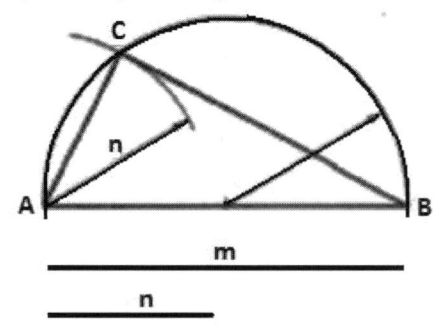

Figure 5-48: *Constructing a right triangle*

To inscribe an equilateral triangle within a circle having a given diameter, AB

Procedure:

1. With center O and radius equal to AM (M is midpoint of AB), draw the circle.
2. With C as center and the same length of radius, draw an arc cutting the circle in points D and E.

3. Join points D, E, and F. The required triangle is DEF.

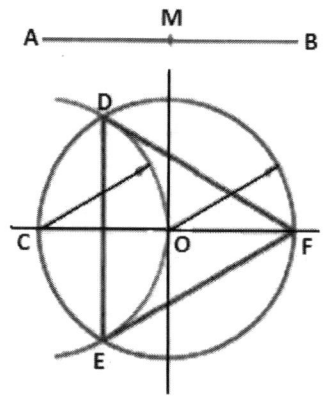

Figure 5-49: Inscribed an equilateral triangle within a circle

5.6 Quadrilaterals, their properties and construction

Any geometrical figure bounded by four straight lines is referred to as a quadrilateral. The straight lines joining opposite angles of the quadrilateral are termed diagonals.

Examples includes

1. *Rectangle*: This has pair of opposite sides equal and parallel. The diagonals are equal and bisect each other at the center. The opposite angles of the diagonals are equal.

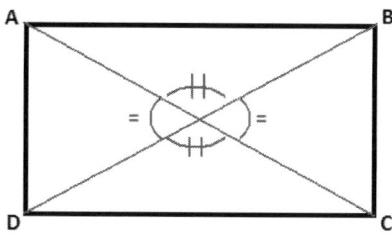

Figure 5-50: Rectangle

2. A *square* has all the four sides equal, the diagonals are equal and bisect each other at 90. The angles the diagonals made to the base are equal.

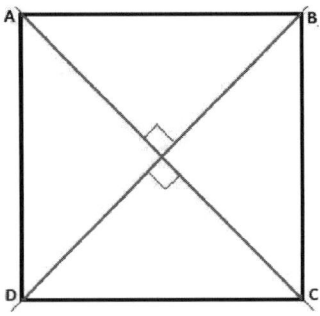

Figure 5-51: Square

3. *Rhombus* is a quadrilateral which has all its sides equal, opposite angles equal. The opposite sides are parallel to each other. The diagonals are not equal but bisects each other at 90

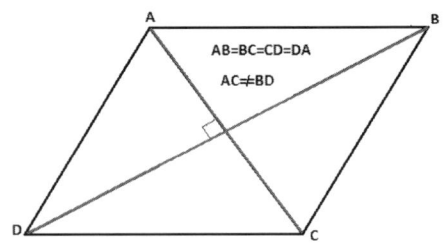

AB=BC=CD=DA
AC≠BD

Figure 5-52: Rhombus

4. *Kite* has two adjacent sides equal, the diagonals bisect each other at 90 and two opposite sides are parallel to each other.

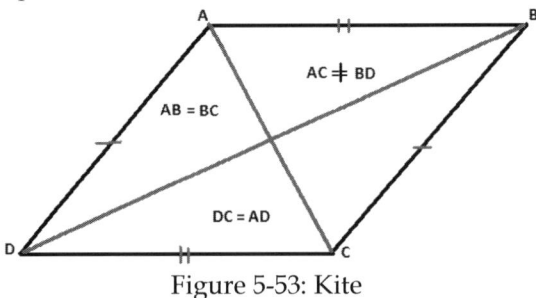

AC ≠ BD
AB = BC
DC = AD

Figure 5-53: Kite

5. *Trapezium* has two of its four sides opposite and parallel, none of its sides are equal and the diagonals are not equal.

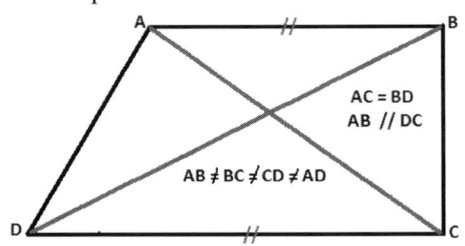

AC = BD
AB // DC
AB ≠ BC ≠ CD ≠ AD

Figure 5-54: Trapezium

6. *Trapezoid* has none of its sides and angles equal.

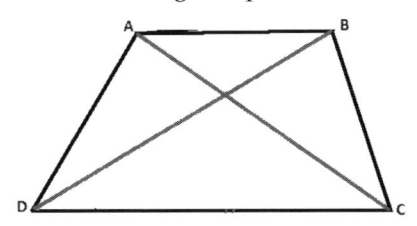

Figure 5-55: Trapezoid

7. *Parallelogram:* This is a quadrilateral having its opposite sides parallel. Examples include rhombus, square etc.

Construction of quadrilaterals

To be able to construct a quadrilateral, it is essential to know either the lengths of the diagonals or the length of the sides in addition to the angles before it can be constructed. The only exceptions to these conditions are the rhombus and the square which are special kinds of parallelograms.

When a quadrilateral is to be measured and constructed, the number of measurements required is five as follows:

1. Four sides and one angle, or
2. Four sides and a diagonal, or
3. Three sides and two angles.

To construct a rectangle on a given base

Procedure

1. Draw the base line AB
2. From points A and B erect perpendiculars.
3. With A and B as centers and radii equal to the perpendicular height describe two arcs to cut the perpendiculars in points C and D.
4. Join CD to complete the figure

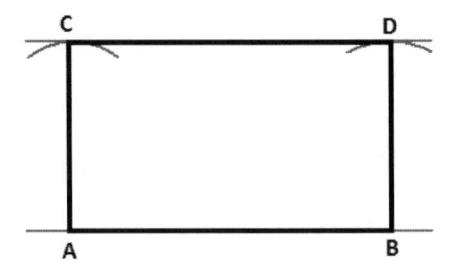

Figure 5-56: Construction of rectangle

Division of quadrilateral into equal parts

To divide a given quadrilateral ABCD into two equal areas by a line which passes through one of the corners, A, proceed as follows.

1. Draw line BD and locate its midpoint, M.
2. Draw a line through point M parallel to diagonal AC to intersect side BC at point E.
3. Line AE divides the quadrilateral into equal areas AEB and AECD.

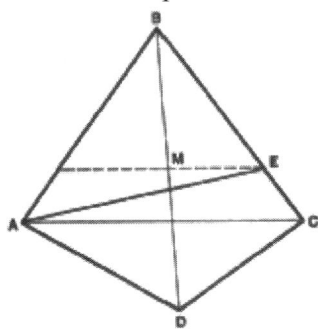

Figure 5-57: Division of quadrilateral into equal area

5.7 Polygon, their properties and construction

Polygons are geometric figures with more than four sides. The first member of this geometrical figure is called pentagon (five-sided polygon). Other members in increasing order includes: hexagon, heptagon, octagon, nonagon and octagon.

Polygons could either be regular or irregular. Regular polygons are characterized by equal lengths and angles, while the irregular polygons have unequal lengths and the angles are not equal.

Construction of regular polygon

Polygons could be constructed in two ways: Set square method and construction method. Set square method utilizes the combination of set squares.

Construction of regular polygon on a given line using the set squares

To describe a regular hexagon on a given line using 60 set squares only, proceed as follows:

1. Draw the given line AB and set up the perpendicular AC
2. From B draw BC at 60 to AB
3. Points D, E, and F are found in a similar way using the 60 set-square to complete the figure.

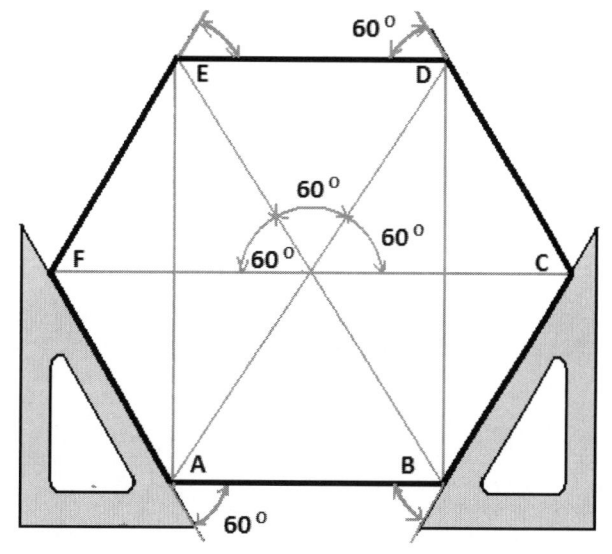

Figure 5-58: Construction of regular polygon using set squares

Construction of polygon on a given line

1. Draw the given line AB
2. Extend the line to C and draw the semi-circle CB with radius CA or AB
3. Divide the semicircle into five equal parts and join A2
4. Bisect AB and A2 to give point O
5. Describe the circle with center O touching points BA2
6. Step off sides DE and EB equal to A2. ABED2 is the required polygon
7. This method is used to construct any regular polygon on a given line.

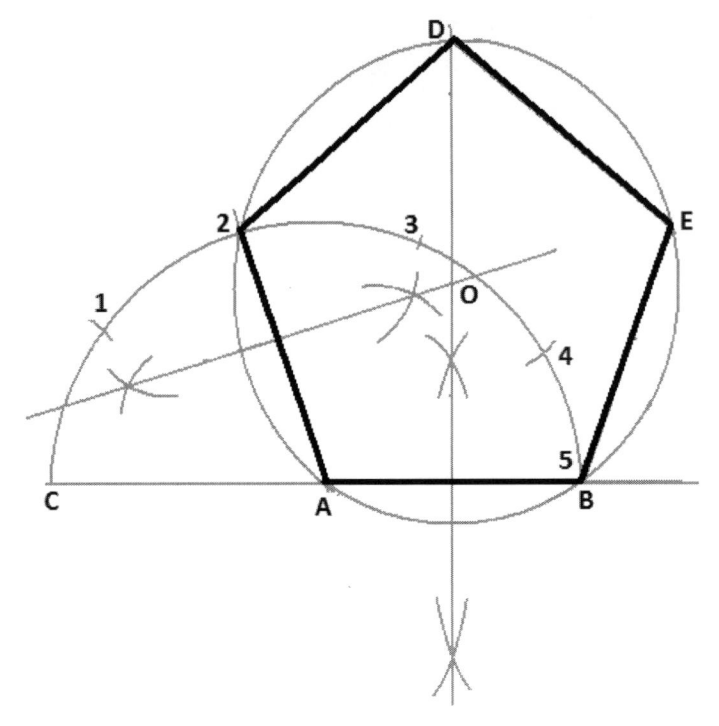

Figure 5-59: Construction of regular polygon using compass

Construction of a regular polygon within a given circle (Heptagon)

1. Draw the given circle of diameter AB in which a regular heptagon is to be constructed.
2. Divide AB into seven equal parts
3. With A and B as centers and radius AB, describe arcs at C
4. From C, draw through the second division to D, then BD become one side of the polygon
5. Step off distance BD seven times round the circle and complete the polygon.

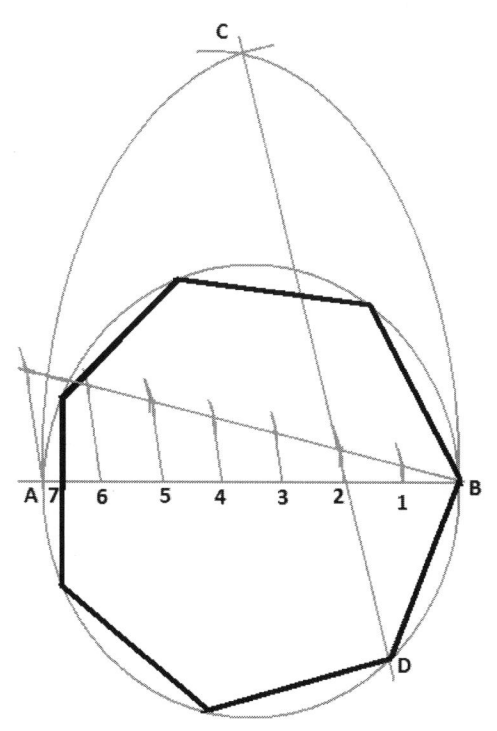

Figure 5-60: Construction of regular polygon using line division

To construct a regular pentagon when the length AB of the sides is known

Procedure

1. First construct BC = AB, and perpendicular to AB.
2. With M (midpoint of AB) as center and MC as radius, draw an arc cutting AB extended at point D.
3. With A as center and AD as radius, draw an arc; and,
4. With B as center and radius BA, draw an intersecting arc to locate point E.

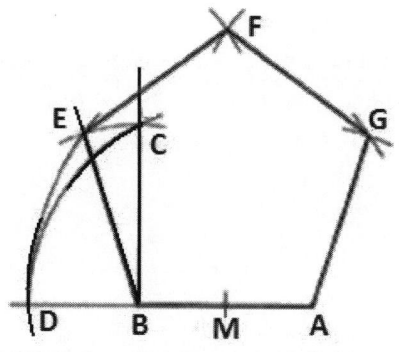

Figure 5-61: Regular pentagon given length of one side

5. Line BE is a side of the regular pentagon. The construction for locating point F and G is fairly obvious.
6. ABEFG is the required pentagon. (The solution of this problem is based upon the fact that the larger segment of a diagonal of the pentagon, when divided in extreme and mean proportion, is the length of a side of the pentagon.

Note that point B divides AD in extreme and mean proportion and that AB is the larger segment of AD.

To construct a regular pentagon when the length EC of a diagonal is known

Procedure

1. First divide EC in extreme and mean proportion (ET squared = EC X TC).
2. With E and C as centers and radius ET, draw arcs that intersect at point D.
3. Draw the perpendicular bisector of ED and locate point O on the vertical line passing through D.

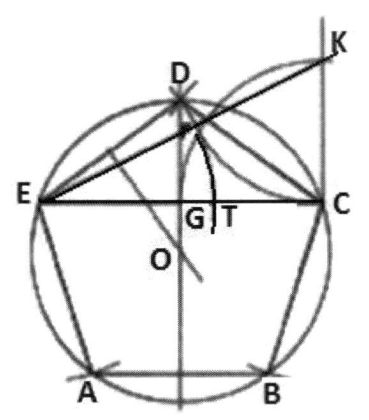

Figure 5-62: Regular pentagon given a diagonal

4. Point O is the center of the circle (radius OE) which circumscribes the pentagon.
5. With E and C as centers and a radius equal to CD (or ED), draw arcs cutting the circle at points A and B.
6. Draw the necessary lines to form pentagon ABCDE.

To inscribe a hexagon within a given circle

Procedure

1. With A and D as centers and a radius equal to the radius of the circle, draw arcs which intersect the given circle in points B, F, C, and E.

2. The required hexagon is ABCDEF.

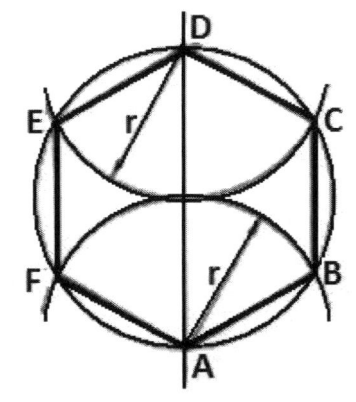

Figure 5-63: Regular pentagon given a diagonal

An alternative construction is shown below.

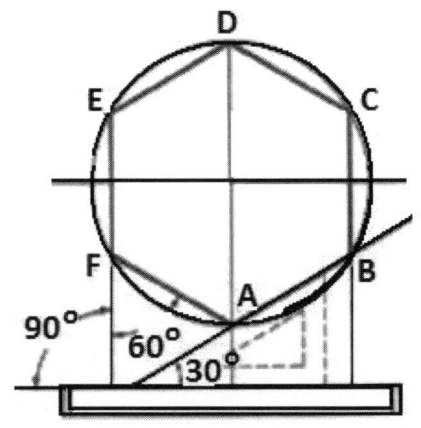

Figure 5-64: Regular pentagon given a diagonal

To construct a regular polygon having n sides

The polygon in this example is a nonagon (nine equal sides) and AB is the given length of each side.

Procedure

1. With B as center and AB as radius, describe a semicircle, and by trial divide it into nine equal parts.
2. Starting from point T, locate the second division mark, C.
3. Locate point O, the center of the circumscribing circle (This is easily done by finding the intersection of the perpendicular bisectors of AB and BC.).
4. Draw the circle with center O and radius OA and complete the nonagon

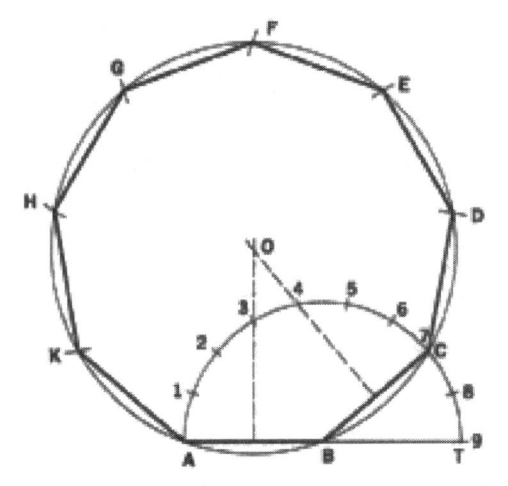

Figure 5-65: Construction of regular polygon using line division

Alternatively, these points are the centers of circumscribing circles of polygons having sides equal to AB

General method of constructing any number of regular polygons

1. Let AB be the given line, bisect the line to determine the center line
2. Erect a perpendicular from B equal to AB
3. Join AC cutting the center line in point 4

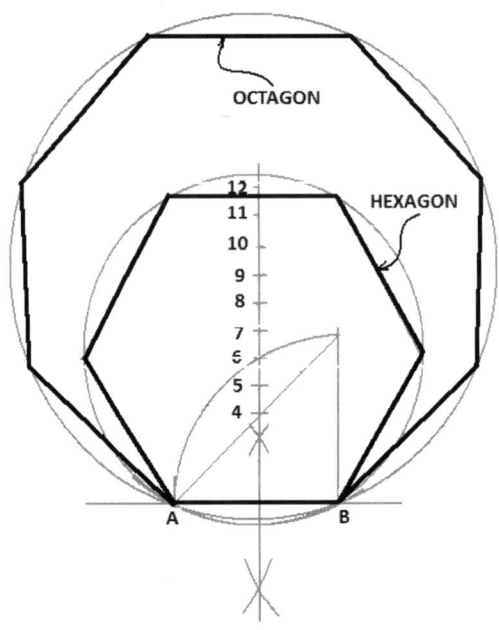

Figure 5-66: Construction of regular polygon using line division

6. With center B and radius AB, draw the quadrant AC cutting the center line at point 6
7. Halve the distance between the point 4 and 6 to give point 5 and step these distance along the center line to give the points 7, 8, 9 etc.

5.8 Circles, their properties and construction

A circle can be said to be the curve that is the locus of points in a plane with equal distance (radius) from a fixed point (center). In elementary mathematics, circle often refers to that finite portion of the plane bounded by a curve (circumference) all points of which are equidistant from a fixed point of the plane.

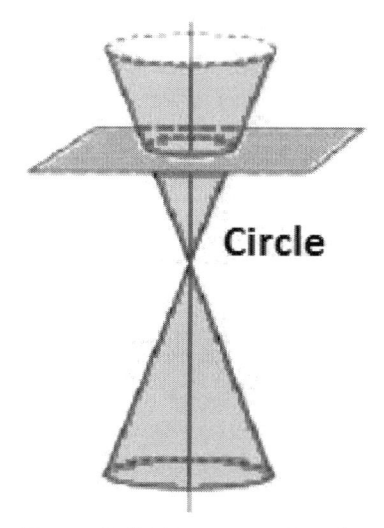

Figure 5-67: Features of circle

Properties of a circle

- *Diameter* – This is the line drawn from any one point P on the circle to another point r passing through the center O. The diameter equals twice radius.
- *Chord* – This is a line segment whose end points x and y are points on a circle. A chord subtends two (conjugate) arcs of the circle. The chord of maximum length passes through the center of the circle and is called the diameter.
- *Arc* – This is either of the two parts of a circle (considered as a curve) intercepted between two distinct radii *aO* and *Or*

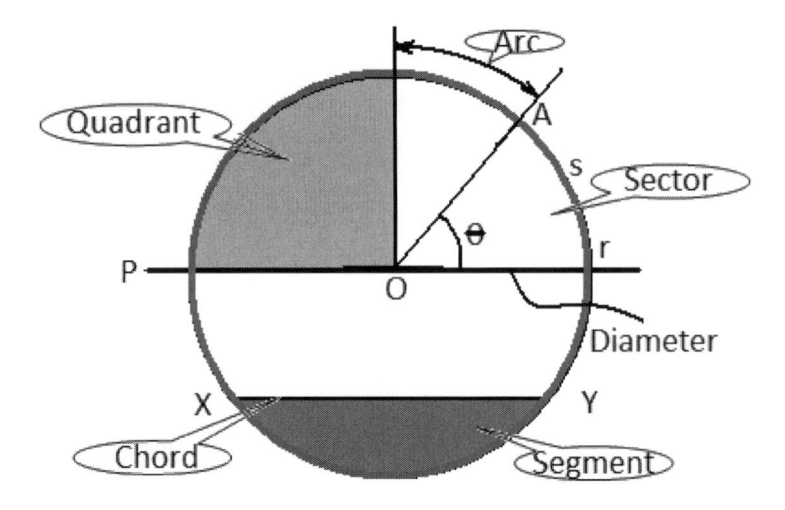

Figure 5-68: Properties of circle

- *Major segment* (Longest distance from x through p to y and x) – This is the largest part of the circle divided by a chord.
- *Minor segment* (Shaded portion in Figure above) – This is the smallest section of the two conjugate arcs of the circle.
- *Circumference* – This is the distance round the circle.

Types of circles

1. *Inscribed circle*

An inscribed circle is one in which a circle is drawn within another bigger circle or plane figure such as triangle or rectangle with all the sides touching the figure internally. Note the circle must not by any means cut through the side or line or the circle failed to touch the line. These are defects which often resulted from inaccuracies in bisection or measuring accurately the given lengths of the figure.

A basic procedure for the construction of an inscribed circle

The basic procedures are as listed below:

a. Draw, by construction the figure in which a circle is to be inscribed. (This can either be triangle, polygon or quadrilateral. But in this case a triangle ABC).
b. By the method of bisection of angles, bisect two or all of the angles as shown in the figure.

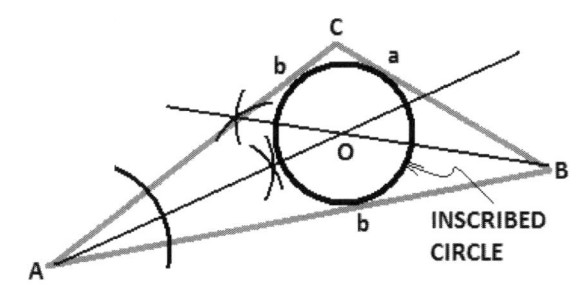

Figure 5-69: Construction of inscribed circle

c. Join the points of intersection of the arcs to their respective angles. Note that the lines meet at a point o.
d. With O as a center and either oa, ob, and oc, as radius, describes the required inscribed circle.
e. Note; it two of the three sides touches the circle respective the procedure to find out where you made mistake.

2. *Circumscribed circle*

A circumscribed circle is that in which a circle is drawn outside the figure. The figure is entirely inscribed in the circle.

Basic procedure for the construction of a circumscribed circle

a. Draw the figure by construction (triangle)
b. By the method of bisection of lines, bisect two of the three lines with the intersecting at a point O. Note: this point of intersection can fall either written or outside the figure.

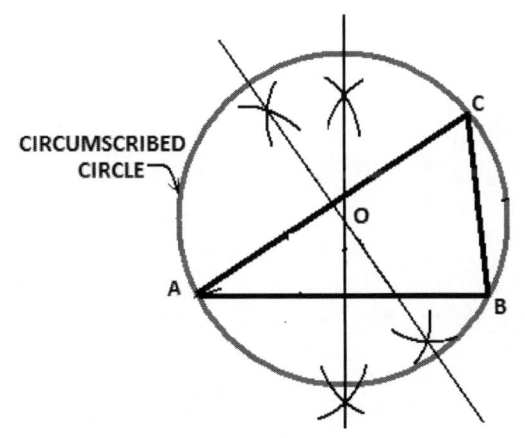

Figure 5-70: Construction of inscribed circle

c. With *o* as a center and *oa* or *ob* or *oc* as a center describe a circle round the figure. That is the required circle.

3. *Escribed circle*

An escribed circle is the circle that is touching a figure (e.g. a triangle) externally. The circle is touching all the sides externally.

Basic procedure for the construction of an *escribed circle*

a. Draw by construction the required figure (triangle) *abc*.
b. Produce line *ac* drawn to a point (Convenient) in space.
c. Produce line *ab* drawn to a point (Convenient) in space
d. Using method of bisection of angles, bisect the external angles at *b* and *c* as shown in the figure 5-62 below.

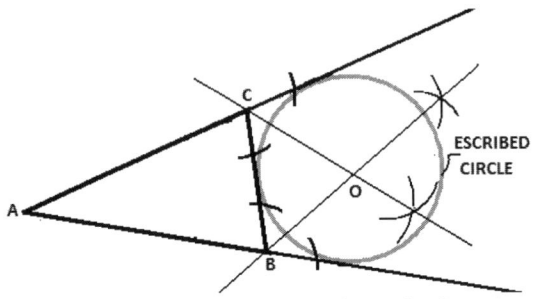

Figure 5-71: Construction of escribed circle

e. The lines of bisections intersect at a point *o*.
f. With *o* as center, extend your compass to touch line *CB*, *AC* and *ab* produced. Then describe the circle.

5.8.1 Circle geometry

When circles touch *internally*, the straight which joins the centers, being produced, will pass through the point of contact (P.C.). See Figure 5-72 below:

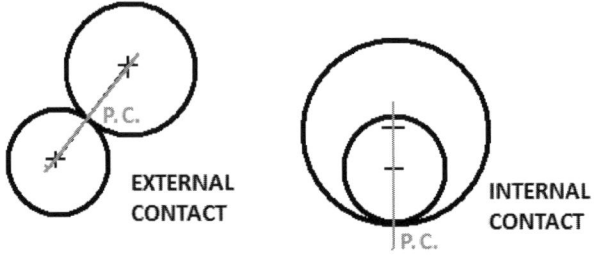

Figure 5-72: External and internal contact

Circles touching lines

The center of a circle moving along a lie describes a line parallel to the given line.

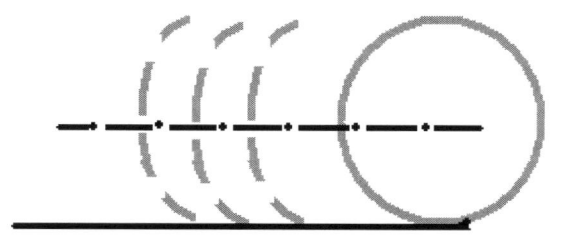

Figure 5-73: Circles touching line

The center of a circle which touches two lines at right angles will be at the intersection of the two parallel lines.

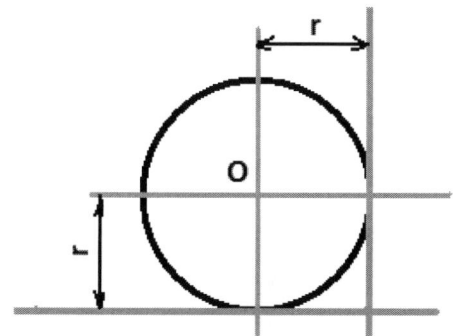

Figure 5-74: Center of circle touching line

Usually in practical drawings only a quarter of the circle is required. Line in the radius first, then finally line in from the points of contact (Figure 5-75).The same method is used to find the center of a circle touching lines at an angle as shown in *step 1*.

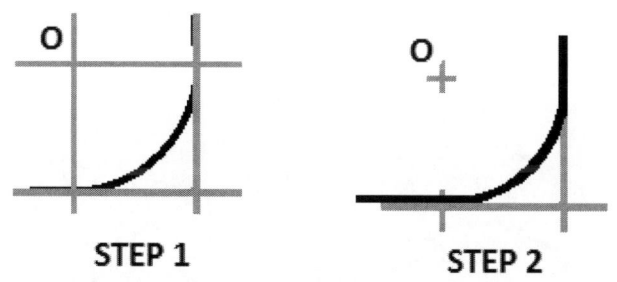

STEP 1 **STEP 2**

Figure 5-75: Drawing the tangent of circle touching line

The points of contact are marked from the center of the circle at right-angles to the lines as indicated in *step 2*. Line-in the radius exactly to the points of contact, *step 3*. Finally line-in from the points of contact of *step 4*.

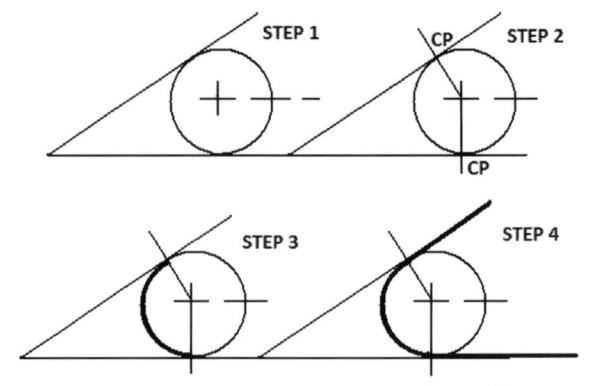

Figure 5-76: Drawing tangent to form elbow

Circles touching circles externally

Circles touching circles externally are best dawn by adding the radii of both circles to determine the other center *o*. To draw a circle radius *r* touching another circle externally, mark the center *o* and draw the circle *r*. To get the other circle add *r* to *r* (R + r). With center *o* and radius R + r describe arc. Draw a line from *o* to any point on the arc. Mark the intersection *p* with *p* as a center and radius *r*, describes the circle.

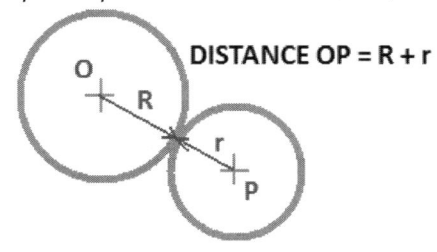

Figure 5-77: Circles touching externally

In the same way, to draw three circles that are touching each other, given the sizes, such as 26mm, 5mm, and 20mm radius as example, first draw the larger circle. The centers of the other circles are then found by adding the radii together. Join the center by lines to find the points of contact.

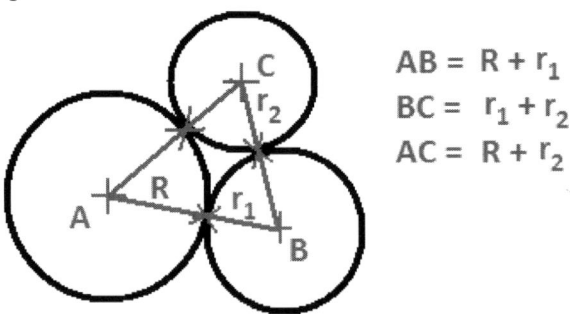

$AB = R + r_1$
$BC = r_1 + r_2$
$AC = R + r_2$

Figure 5-78: Measuring distance of circles touching externally

When a circle touches another circle *externally* the straight line which joins the centers passes through the point of contact.

Circles touching circles internally.

The same procedure as above is followed only that the other circle centers are obtained by subtracting the smaller radius from the larger (R-r).

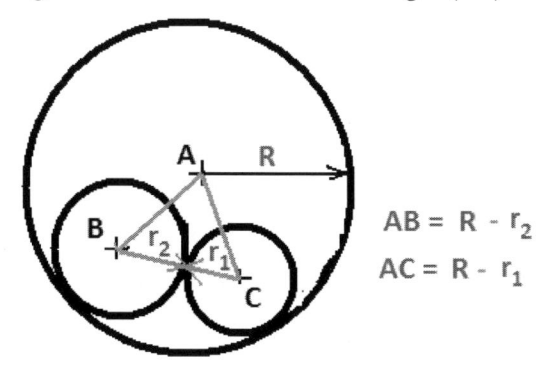

$AB = R - r_2$
$AC = R - r_1$

Figure 5-79: Circles touching internally

Circles touching circles and lines

This condition is best described when the radius or diameter of the circle and the distance from the circle to the line are given.

a. Given the radius of the circle as R, and a distance d, from the line, draw a circle radius r touching the circle and the line AB.
b. Draw the line and measure the distance d.
c. Describe the circle radius R from point d as center and radius R + r.
d. Describe an arc from the point on line AB, and describe another arc cut the already produced arc at C.
e. With center C, describe the circle to touch the circle radius R and the line AB.

DISTANCE OC = R + r

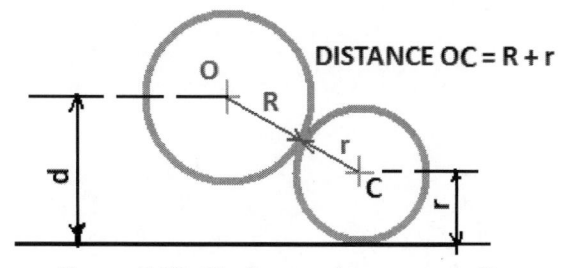

Figure 5-80: Circles touching externally

For circles touching a circle and a line internally, the procedure for construction remain the same as above but the distance OC = (R − r). You can try practicing this method.

Circles touching circles and points

Given a circle radius and point distance from circle center, circles can be drawn to touch the circle and the point. Given the circle radius *r*, center and point *a* distance *d* (*x, y*) from center *o*. To draw an external circle radius r touching the circle radius *r* and point *a*, follow these procedures.

1. Draw the given circle radius *r*.
2. Locate the point *a* distance *d* (*x, y*) from center *o*.
3. Locate the other center *c* radius *r*.
4. With R + r, describe an arc.
5. From point *a* as center and radius *r* locate center *c* at the point of intersection of the arcs.
6. Describe the circle radius *r* to touch the other circle and point *a*.

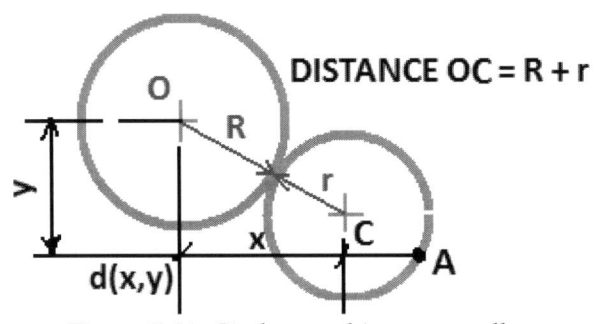

Figure 5-81: Circles touching externally

In the same way and procedure, draw a circle touching circle and points internally.

To draw an arc of given radius, r, tangent to two given lines m and n

The construction is clearly shown in the figure.

Procedure

1. Draw the given lines m and n.
2. Bisect lines m and n and draw parallel lines m′ and n′ at distance r (radius of the arc) to meet at point O

3. The center O of the required arc is at the intersection of lines m' and n', which are parallel to lines m and n, respectively, at distance r.

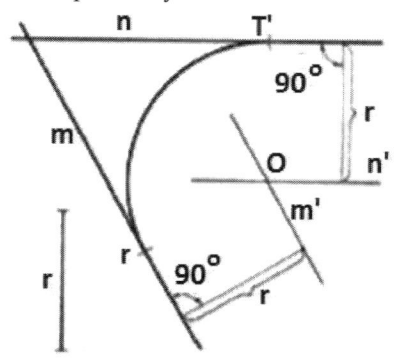

Figure 5-82: Drawing tangents to two given lines

To draw arcs of radius r, tangent to a given line, m, and a given circle, O

Procedure

With O as center and radius (R + r), draw an arc cutting line m' (line m' is parallel to m at distance r) at points P and Q, which are the centers of the required arcs.

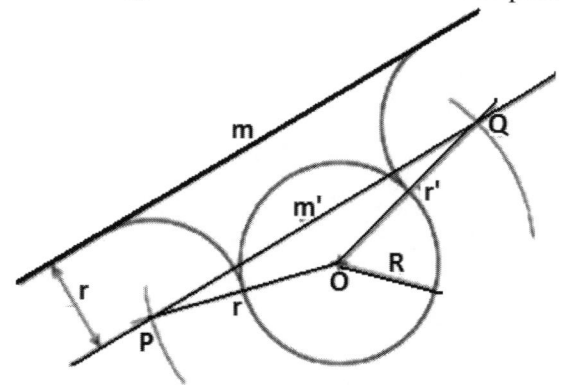

Figure 5-83: Drawing arcs, tangents to given line and circle

To draw arcs of radius r, tangent to two given circles

Procedure

With center O and radius (R + r), draw an arc; and, with center O' and radius (R' + r), draw an arc. The intersections of the two arcs are P and Q, which are the centers of the required arcs.

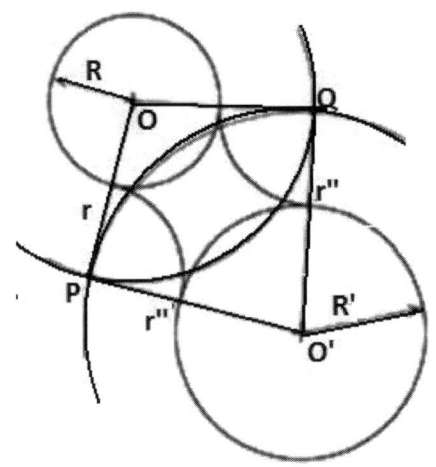

Figure 5-84: Drawing arcs, tangent to two given circles

To divide a circle into seven equal parts by concentric circles

Procedure:

1. First, draw a semicircle on OA as diameter.
2. Then divide OA into seven equal parts and construct verticals to intersect the semicircle in points 1, 2, etc.
3. Finally, draw the required concentric circles with radii 0-1, 0-2, 0-3, etc.

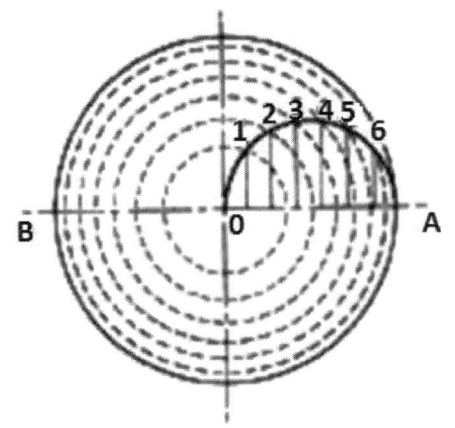

Figure 5-85: Dividing circle into equal parts by concentric circles

Bisection of chord

Let's briefly discuss the bisection of a chord before going further. A straight line *cd* which bisects a chord *ab* of a circle at right-angles passes through the center *o* of the circle.

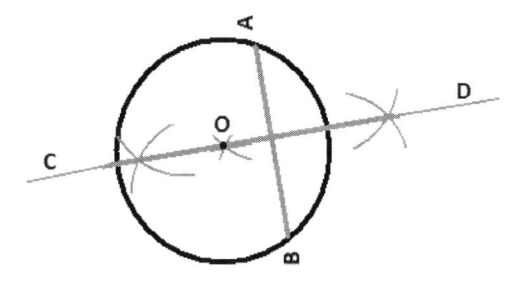

Figure 5-86: Bisection of chord

5.8.2 Tangency and normalcy construction

The straight line which is drawn at right angles to the diameter of a circle from its extremity is called a *tangent*. A line drawn perpendicular to a tangent is called *normal*. Therefore tangency and normalcy are the principles of drawing tangents and normal to any given circle at any given point.

External tangents

Tangents can be constructed externally to circles as the case may be, most important, the procedure to follow is the same.

   a. Draw the two circles and establish the centers O and D.
   b. Join the two centers of the circles.

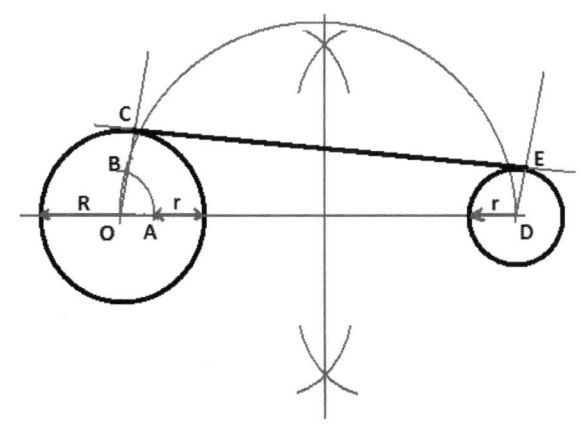

Figure 5-87: Tangent to circles touching externally

   c. Bisect OD and draw the semicircle with diameter OD.
   d. Subtract the radius of the small circle from the larger circle to obtain radius OA.
   e. With center O, radius OA, draw an arc to cut the semicircle at point B.
   f. Join O to B and produce a line to cut the circumference at C.

g. C is the point of contact, and DE is drawn parallel to OC join CE as the tangent.

Internal tangents

In the same way as drawing the external tangent, follow the following procedures:

1. Draw the semicircle
2. Add the radius (T + r) of the bigger and smaller circle.

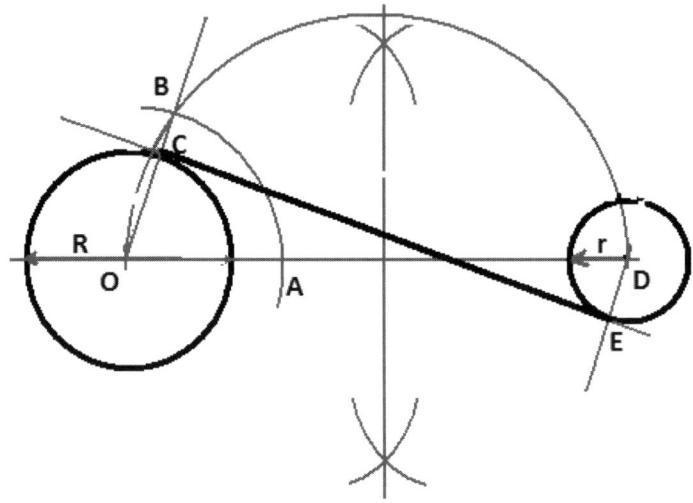

Figure 5-88: Tangent to circles touching internally

3. Draw an arc radius OA, center O to cut the semicircle at B.
4. Join O to B. the point of contact is at C where the line OB cuts the circumference.
5. DE is drawn parallel to OC.

Construction of tangents

1. *Construction of tangent to a circle from a point outside (space)*

If required to draw a tangent to a circle from a point anywhere outside the circle.

Step 1: Draw the required circle diameter d, locater the point *p* anywhere outside the circle.

Step 2: Join *p* to the center of the circle *o*. Bisect *op* and draw the semicircle with *op* as diameter. Mark point *c* where the semicircle cuts the diameter circle. This is the point of contact. Join point *p* to *c* that the required tangent.

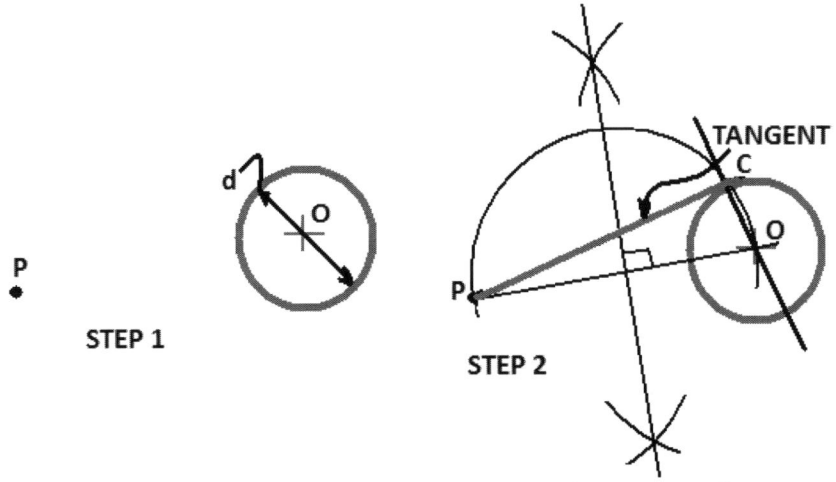

Figure 5-89: Construction of tangent touching externally

2. *Tangent to a circle from a point on the circumference*

To draw a tangent to the diameter of a circle from a point *a* anywhere on the circumference, Draw the normal from the center through point *a* located anywhere on the circumference. Construct the tangent at right angles to the normal through point *a*.

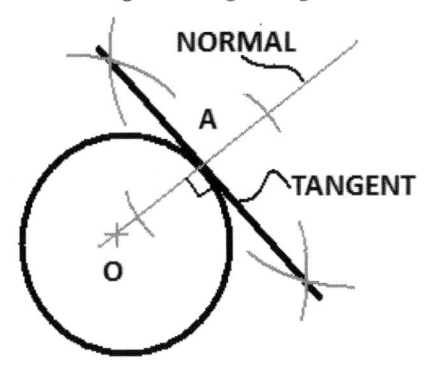

Figure 5-90: External tangent

3. *Tangents to two circles*

A tangent which is drawn to two similar circles will be at right-angles to the centers of both circles.

With two equal circles (circles with equal diameters) the tangent, radii and distance between the centers form the sides of a rectangle.

TANGENT

CENTER LINE

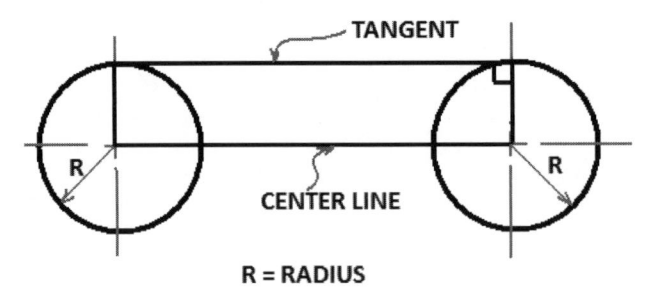

R = RADIUS

Figure 5-91: Circles touching externally

The points of contact between a tangent and unequal circles can be found by constructing a rectangle with smaller radius.

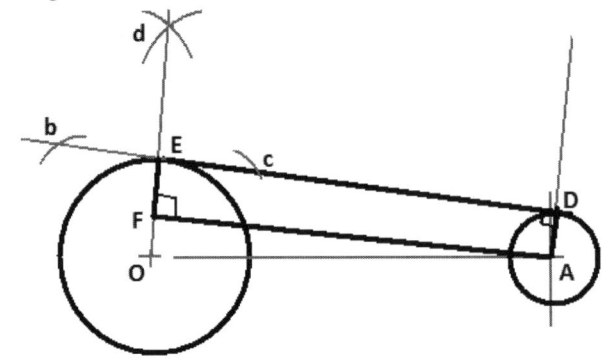

Figure 5-92: Circles touching externally

Procedure:

1. With your compass, draw the two circles and establish the center distance *oa* between them.
2. With center *o* and radius *ob* or *oc* draw arcs intersecting the tangent *ed* at points *b* and *c*
3. With centers b and c and convenient radius, draw two arc to intersect at point d. join line *od*, *e* is the point of contact of the circle with the tangent.
4. With radius *ad* measured on *oe*, complete the rectangle *adef*
5. Points *e* and *d*, are the required points of contact

To draw tangents to two given circles

Procedure:

1. Draw line O-O', which joins the centers of the circles,
2. Layoff from O', distance O'B = (R-r), and O'C = (R + r).

3. With O' as center and radii O'B and O'C, draw arcs which intersect the semicircle on O-O' at points D and E, respectively.
4. Draw O'D to intersect circle O' at F, and similarly draw O'E to intersect circle O' at G.
5. Through center O draw OH parallel to O'F and OK parallel to O'G.
6. Lines FH and GK are tangent to both circles.

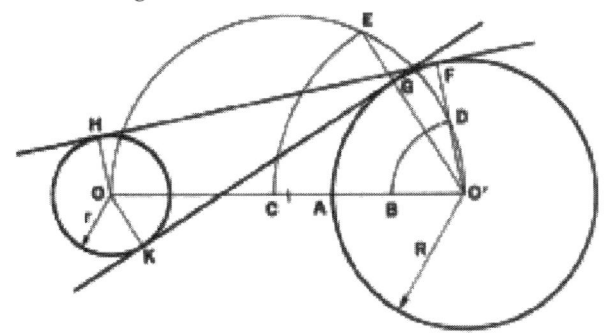

Figure 5-93: Circles touching externally

Exercise

Geometrical figure

1. Describe four different forms of lines and sketch them
2. How should a point be represented on a drawing?
3. Describe the areas of application of the following lines: Construction line, ii. Finished detail lines, iii. Center lines, iv. Projection lines, and v. Hidden detailed lines
4. Define the following geometrical figures and sketch them: a point, line, circle, triangle, rectangle, polygon and quadrilaterals
5. Describe the following types of angle giving graphical illustrations: acute angle obtuse angle, right angle, reflex angle, complimentary angle, and supplementary angle
6. What are the general properties of quadrilaterals? Give examples of quadrilaterals

Construction of triangles

1. Construct a triangle given the following lengths of side 105mm, 70mm, 80mm. inscribe a circle to touch each side.
2. Construct a triangle given AB = 99mm, angle CAB = $37\frac{1}{2}°$, vertical height = 42mm. about this triangle, circumscribe a circle.

3. Construct a triangle AB = 5.3cm, BC = 36cm, AC= 4.5cm, produce sides AB and AC and escribe a circle to the triangle. Measure and state the diameter of the circle to the nearest millimeter.
4. An end view is given of a metal wall mounted holder for a roll of kitchen paper. Copy the diagram and show by construction the largest diameter of roll which can be placed in the holder and touch both sides and the base.

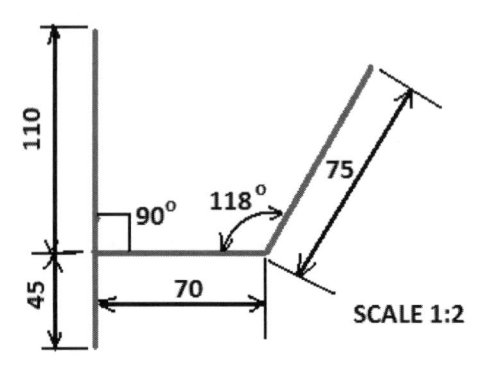

SCALE 1:2

5. Show full geometrical construction of those patterns of 3 and 4 circles of equal diameter within equilateral triangle and a square respectively.

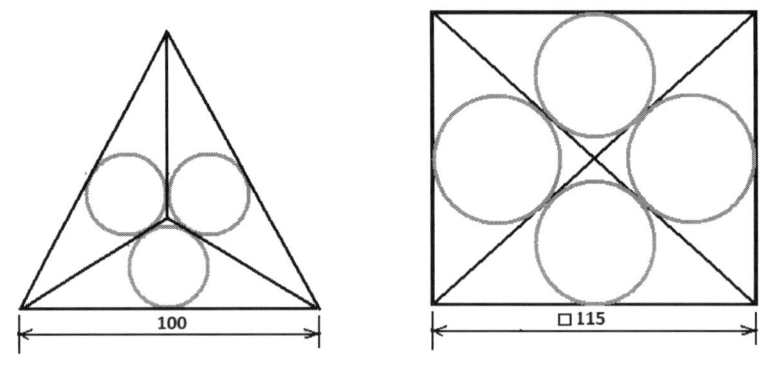

6. Construct an arc with 5/8" radius tangent to arc B and line BD

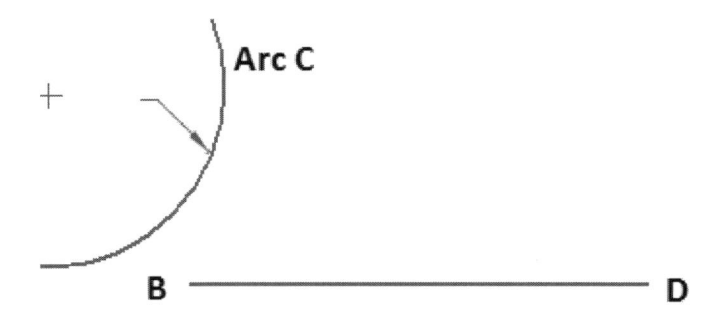

Arc C

B ——————— D

Tangency and normalcy problems

1. The engine crank of an automobile is as shown in the Figure below. Draw to scale (1:2) and show all the construction details in faint outlines.

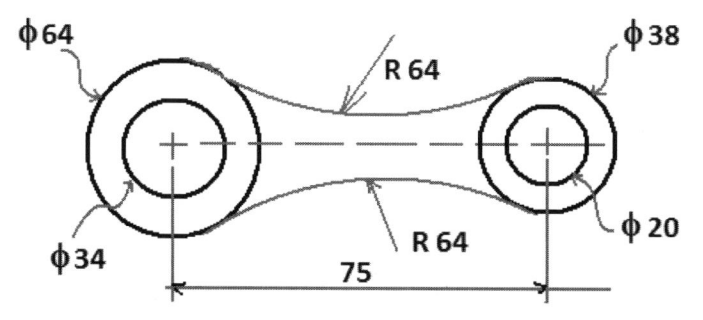

2. The lever area of a lift mechanism is shown this figure. Construct the figure to scale ratio 1:1.

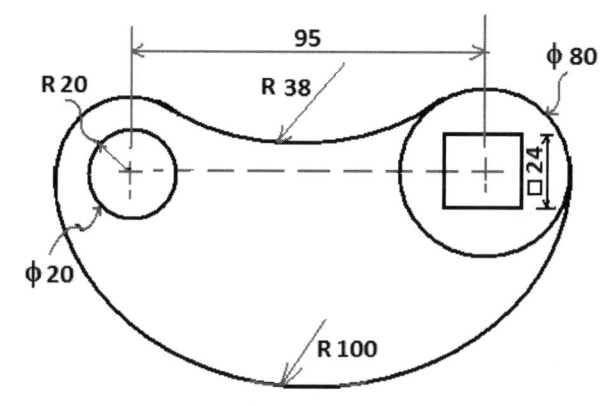

3. Reproduce the figures below by construction. Note that the circle arcs evenly with the lines.

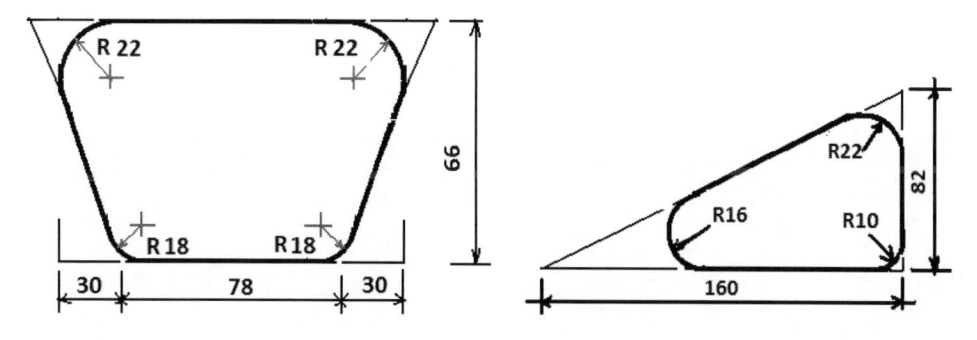

4. The blank for an adjustable spanner is shown below construct to full scale.

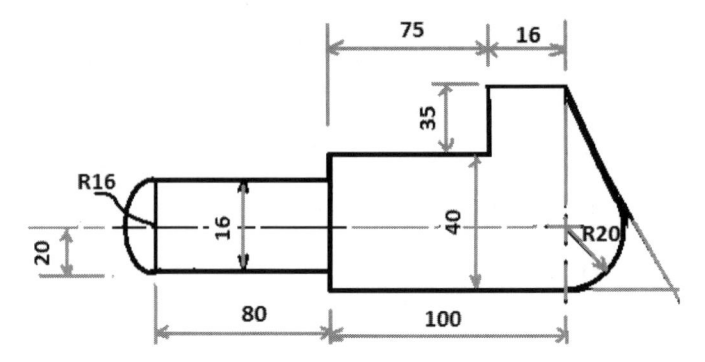

5. The drawing shows a belt drive around two pulleys with an idler pulley between them. Find by construction, the exact points -f contact between the three wheels and the belt.

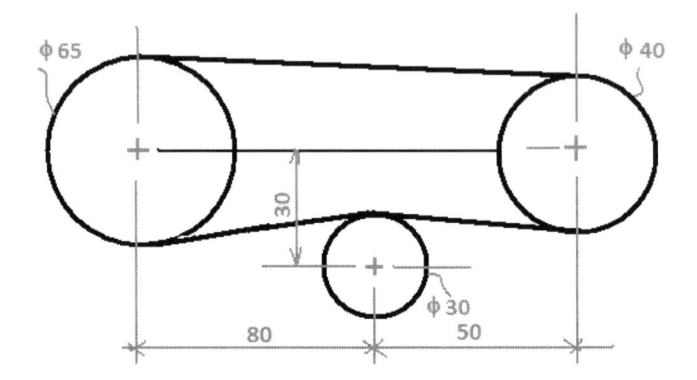

6. The drawing below shows a connecting rod *ab* forming a tangent to the circle through which the crank arm *ao* rotates. Draw the mechanism in this position showing clearly how the tangent is obtained measure and state *ab* to nearest mm.

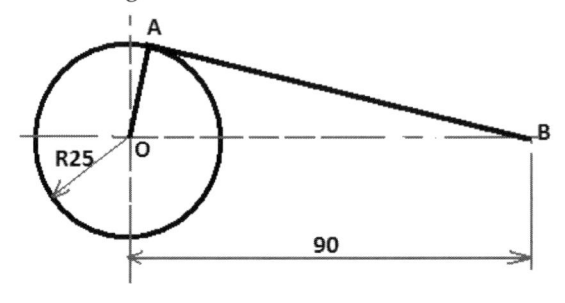

7. Draw the given elevation of a pedestal bearing to size. Full construction must be shown for the lie lined *ab* and *cd* which are tangents. Mark the points of contact with the circle and state the length of the tangents.

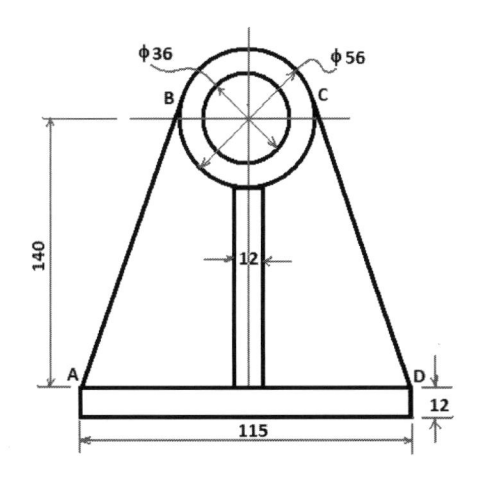

8. Set out the bracket to the given dimensions showing clearly your construction for obtaining the tangential side AB and the R35 radius tangent arc.

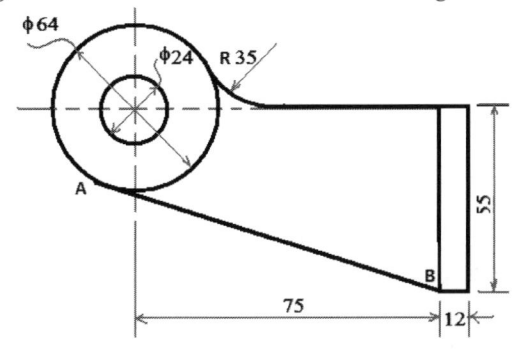

Chapter 6

Projections in Engineering Drawings

6.1 Drawing presentation

Drawings do appear or represented in two basic forms: either as a *plane geometrical figure* or *solid geometrical figure*.

Plane geometrical figure

Plane geometrical figures are figures expressed or represented in two dimensions. Examples include lines, points, triangles rectangles etc., while Plane figures such as squares, triangle, polygon, circle etc. have no thickness but they have areas. Usually they have regular geometrical shapes; hence another name for this plane figure is *lamina*.

To find the shape of these figures in plan and elevation when they are inclined at an angle to a plane, first draw the figure parallel to the plane to find the three sizes before projecting at any angle.

Solid geometrical figure

Solid geometrical figures are in three dimensions (i.e. 3 surfaces are identifiable in one figure which include; front view, side view and plan view). Examples of solid geometrical figures are typically all of the pictorial drawings.

Solid figures: In the projection of a solid objects, perpendicular lines or projectors is drawn from all points on the edges or contours of the object to the plane of projection.

Shown below is the projection of an object onto the frontal plane

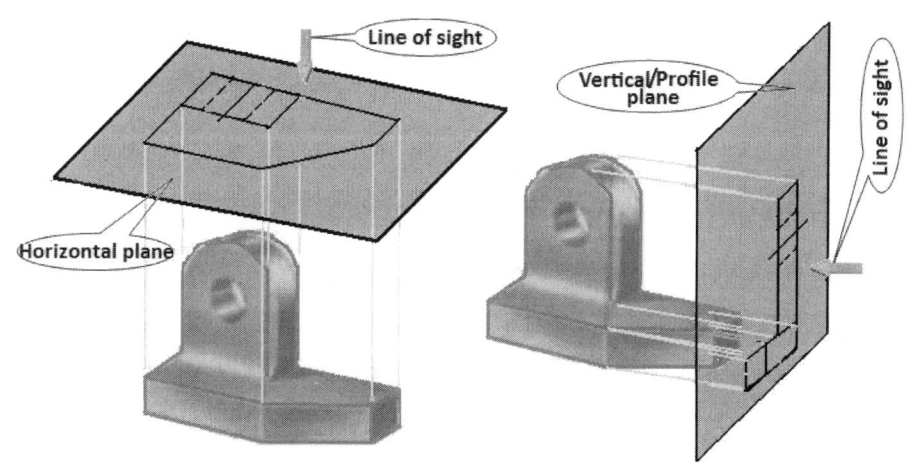

Figure 6-1: Projection of objects

Distinctions between geometrical figures

There are distinctions or differences between these two forms of presentations as listed below:

Plane geometrical figures

Plane geometrical figures could only appear in Multiview projection i.e. one object broken down to many parts such as front view, side view or plan. Multiview projections are a collection of flat 2-D drawings of the different sides of an object. Multiview projections are used to overcome the weaknesses of 3-D projections.

Solid geometrical figures/objects

Solid figure/objects (3-D objects) can be shown as a 3-Dimmensional projection. 3-D projections are useful in that they provide an image that is similar to the image in the designer's minds. 3-D projections are often weak in providing adequate details of the object, and there is often some distortion of the object. For instance, a circular hole becomes an ellipse in an isometric 3-D projection.

The detail studies on projection of these figures are discussed in subsequent chapters.

6.2 Features of projection

Definition of projection

A view or representation of an object is technically known as projection. The following descriptions can be given to fully describe what projection is.

1. Projection can be said to be the representation of the image of any object on a plane or planes such that there is a full description of such object.
2. Projection also refers to the representation of figure or solid on a plane as it would look from a particular direction.
3. A projection is also a view conceived to be drawn or projected on to a plane, known as the *plane of projection*

Projection could either be *descriptive* or *pictorial* as will be discussed in this and subsequent chapters of this book. The goal of all engineering graphics, either with freehand, aided instrument or computer aided design (CAD), is to represent a physical object in engineering drawing forms.

All these drawings are produced and viewed on a surface called a p*lane.*

Planes of projection

Two types of planes are identified in projection drawings as described below;

1. *Principal plane:* This comprises of a horizontal plane (HP), a vertical plane (VP), and the intersection of the two planes. These planes divide space into four quadrants so that the line or object to be represented is found in one of the quadrants. The horizontal plane (HP) produces the *plan* while the vertical plane (VP) produces the *elevation.*

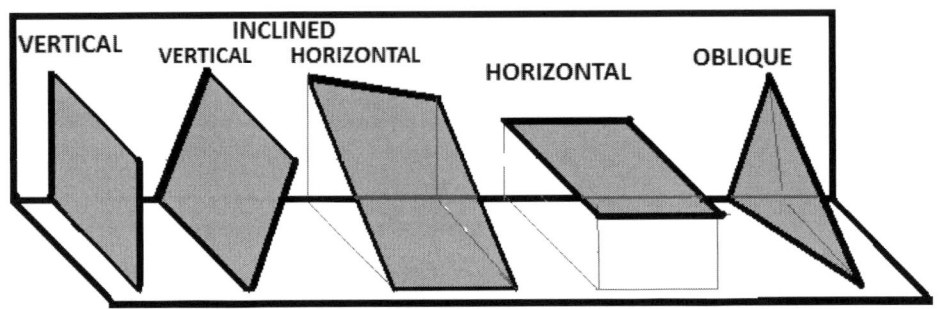

Figure 6-2: Surface projection on principal planes

To show these views on a plane surface, both planes are opened out in the 90° direction of arrow indicating the direction of opening. Four types of planes are identified as described; vertical plane, inclined plane, oblique plane and horizontal plane. The pictorial (surface) projection of these planes used in drawing is represented in the principal planes as shown in Figure 6-1 and their corresponding trace or orthographic views is shown in Figure 6-3.

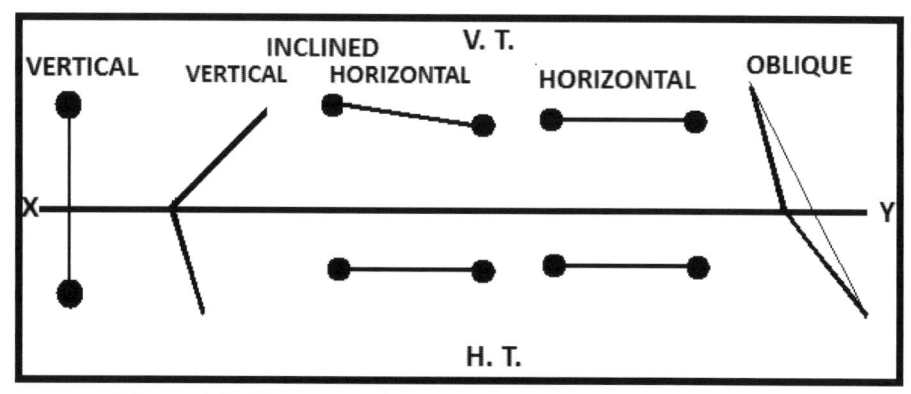

Figure 6-3: Trace or orthographic views on principal planes

2. *Auxiliary plane*: This is a plane of projection other than the two principal planes. The auxiliary plane has no permanent position and thus they are flexible and can be moved as occasion permits.

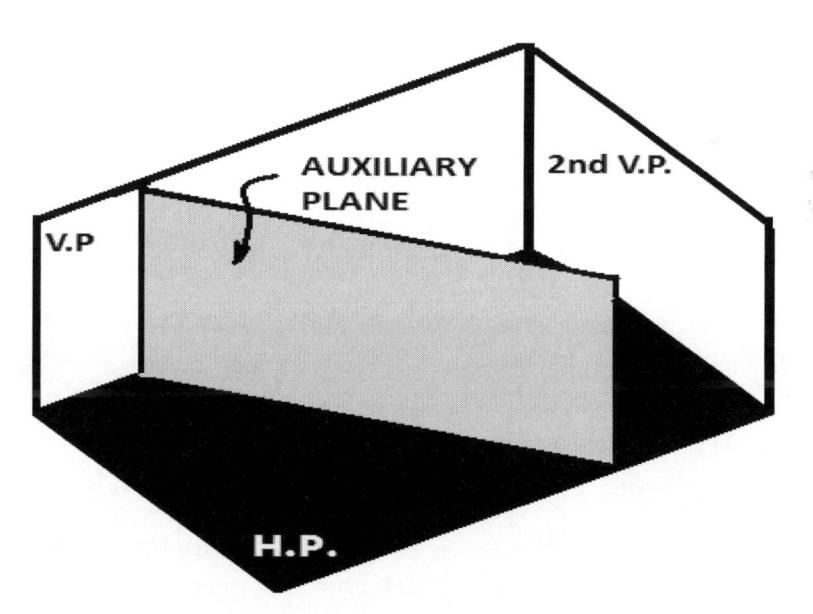

Figure 6-4: Auxiliary plane within the principal planes

6.3 Projection of points

Point projections are carried out in the first quadrant in orthographic views. Two views of a point are represented on the vertical plane and the horizontal plane. The plan is projected on the horizontal plane while the elevation is projected on the vertical plane as shown in Figure 6-5 when the horizontal plane is opened in rabatment, such that the intercession of the two planes were separated by line XY (ground level).

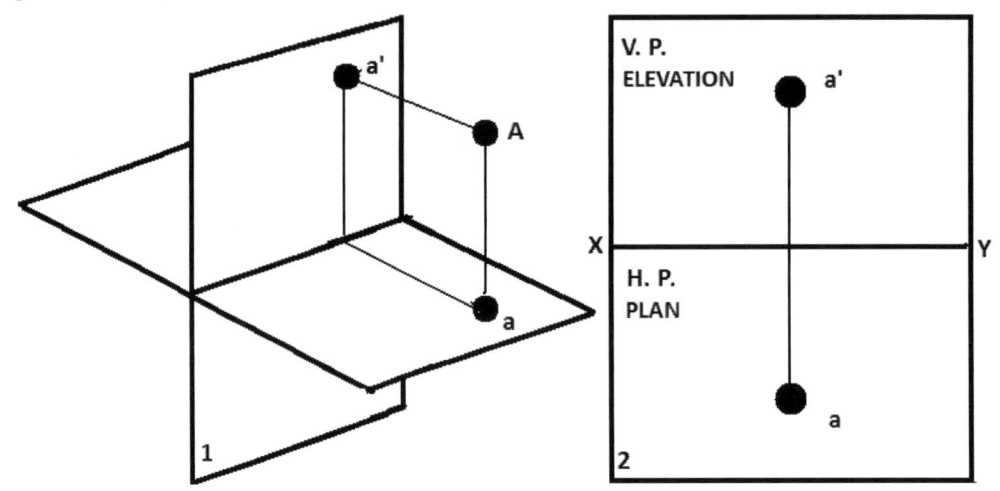

Figure 6-5: Projection and representation of points

6.4 Projection of lines

The great majority of practical works done on drawings is in the 1st quadrant and is represented on the principal planes.

Projection of lines in the principal planes

Projection of lines in the principal planes is vertical, parallel, inclined or oblique to either of the planes in consideration.

Vertical lines: These are lines drawn at 90 degrees or perpendicular to the horizontal plane

Inclined line: A line is inclined when it is parallel to one principal plane only. For instance, in Figure 6-6, figure number four (4) is an example of an inclined line.

Oblique line: A line becomes oblique when it is inclined to both principal planes of projection. Also in Figure 6-6, figure number five (5) is an example of an oblique line

Parallel lines: These are series of lines drawn at equal distance away from each other. They are capable taking any form of orientation in any of the principal planes.

Points and line tracing in the principal planes

Traces of points in drawings often intersect the HP and VP the trace on the HP is called horizontal trace and that on the VP the vertical trace. The projection of lines on that plane gives its true length and inclination.

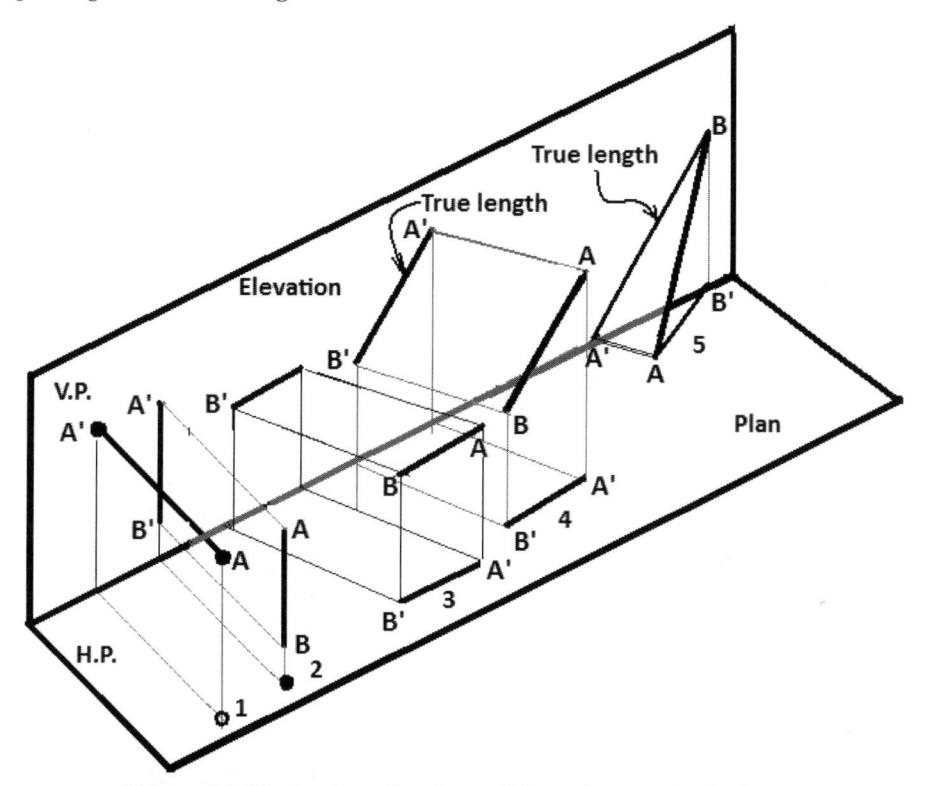

Figure 6-6: Projection of point and lines in principal planes

For instance, Figure 6-6 shows the projection of different points and lines in the 1st quadrant in both the HP and the VP. In the figure, the interpretation of each number from 1-5 is given below:

1. Projection of point A in the 1st quadrant.
2. Projection of line AB perpendicular to the horizontal plane.
3. Projection of line AB parallel to both H.P and V.P.

4. Projection of line AB parallel to the V.P and inclined to the H.P.
5. Projection of line AB inclined to and with the ends in, both the H.P and the V.P.

True lengths and construction

The true lengths of line AB in each of the cases in Figure 6-5 is shown in the plan or elevation only for case 4. An auxiliary plane is required to show the true length of AB in case 5. When a line is parallel to either the HP or VP; the projection of the line on that plane gives its true length and its inclination.

True length of an oblique or inclined line can be found by construction in two different methods.

Method 1: The following procedures are valid:

1. Determine the true position of the line- either inclined to both the vertical plane and horizontal plane or to one of them
2. Draw the straight line in its views on the vertical plane and horizontal plane.

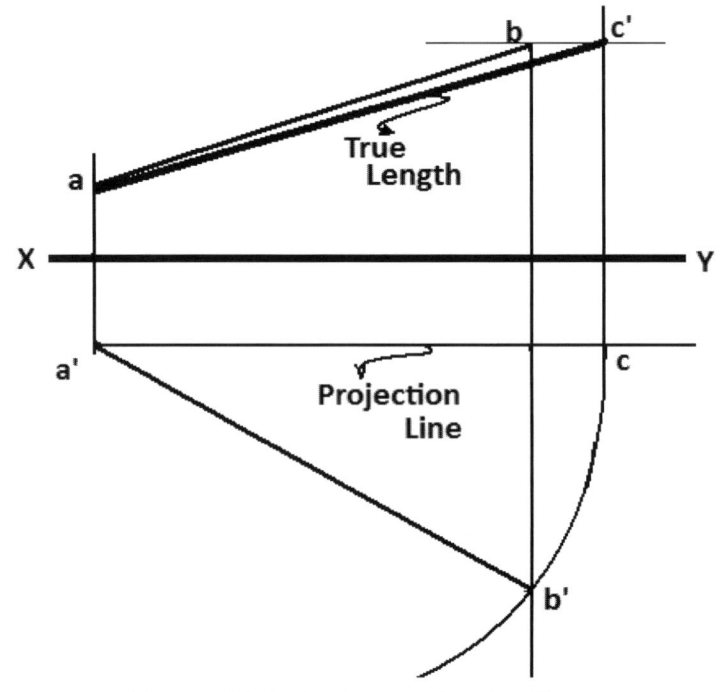

Figure 6-7: Projecting true length of lines

3. With a' as center and a'b' as radius, describe an arc cutting a produced line a parallel to the XY plane at point C

4. Draw a perpendicular from C to cut a projection from b' in c'
5. Join ac' as the true length of the line.

Method 2: The following procedures are valid:

1. Determine the true position of the line ab- either inclined to both the vertical plane and horizontal plane or to one of them
2. Draw the straight line in its views on the vertical plane and horizontal plane.
3. With a' as center and a'b' as radius, describe an arc cutting a produced line a parallel to the XY plane at point C
4. With a pair of compass, center O and radii oc, Ob' and Ob", describe arcs to meet line XY
5. Draw a perpendicular from C', x and y to cut a projection from a and b in a" b" and c' respectively
6. Join ac' as the true length of the line.

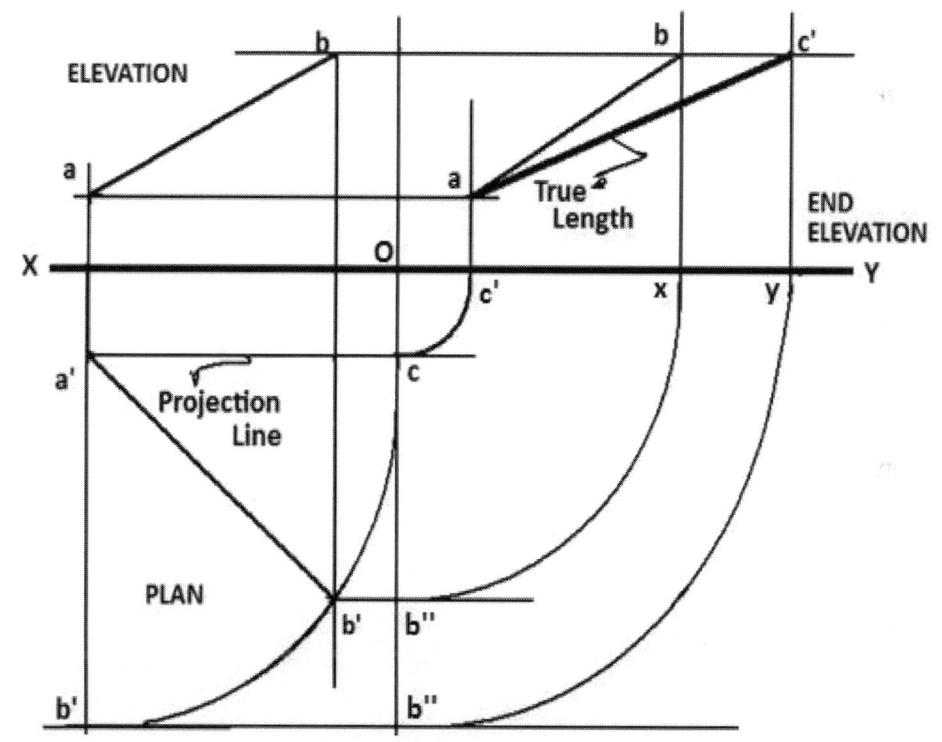

Figure 6-8: Projection of true length of lines

Figure 6-8 below shows a straight line AB in the 1st quadrant inclined to both horizontal and vertical plane; its projections being ab on the vertical plane and a'b' on the horizontal plane.

(a) (b)

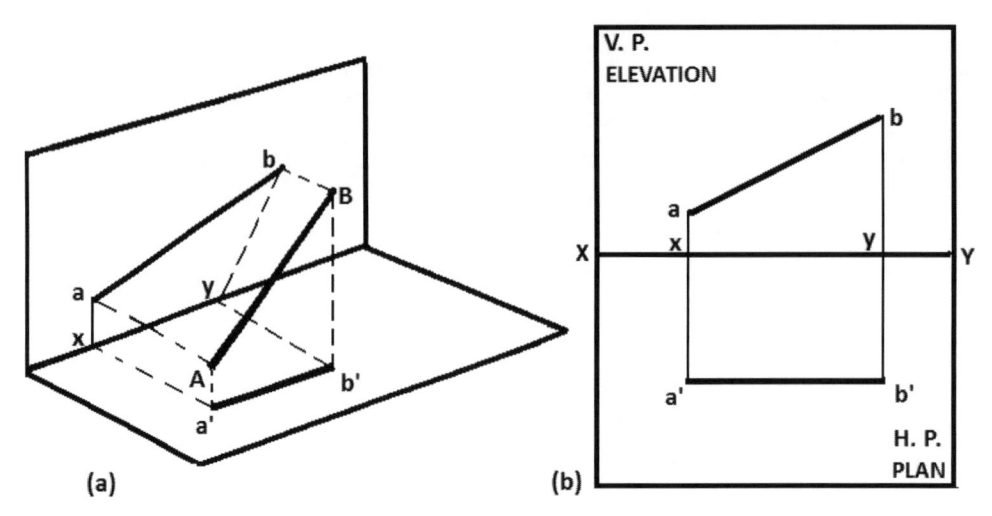

Figure 6-9: Projection of inclined line

The heights of A and B above are shown in Figure 6-9 (b). The heights A and B above the horizontal plane and vertical plane are taken from Figure 6-9 as Aa′ and Bb′

$$Aa' = xa, \; Bb' = By \text{ on the vertical plane}$$

$$Aa = xa, \; Bb = yb' \text{ on the horizontal plane}$$

In Figure 6-8 (b)

$$aa' = ax + xa, \; bb' = yb + yb'$$

Example: Three points A, B and C are shown in plan. Complete the drawing if the elevations of the points are A = 40mm, B = 25mm and C = 30mm form the H.P what is the actual distance between A and B, and B and C?

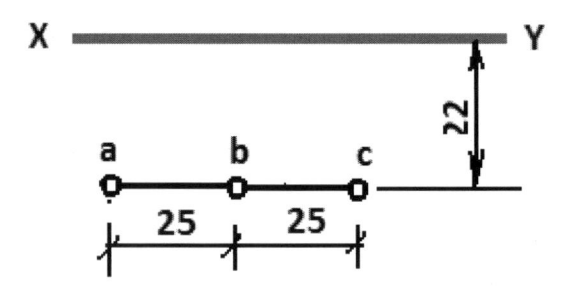

Solution

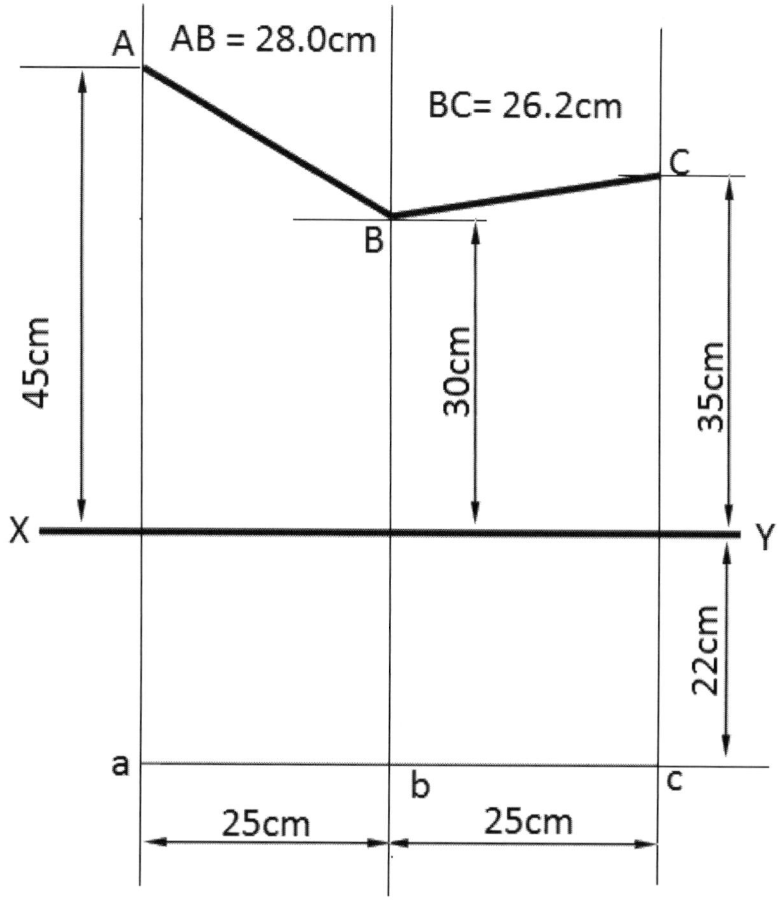
A
AB = 28.0cm
BC= 26.2cm
C
B
45cm
30cm
35cm
X
Y
22cm
a
b
c
25cm
25cm

Example: If AB = 50mm and A is 15mm from HP and 30mm from VP. Point B is 5mm from VH and 35mm from HP. By construction find the length of the line on VP and HP.

Exercise

1. If AB = 50mm and A is 15mm from HP and 30mm from VP. Point B is 5mm from VH and 35mm from HP. By construction find the length of the line on VP and HP.
2. In the following figures, complete the drawings in plane figures and describe in words the precise position of these points A, B, C, and D in relation to the vertical and horizontal panes. Example; point A is – mm from the H.P and – mm from the V. P.

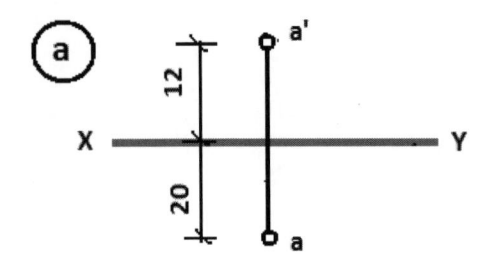

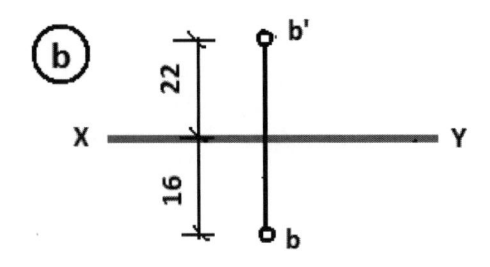

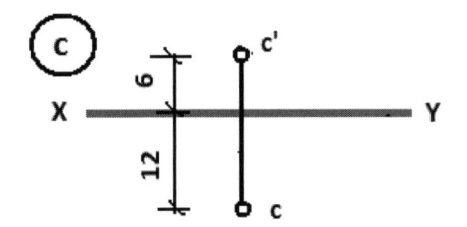

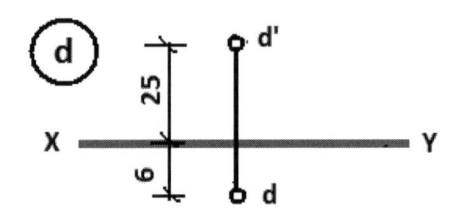

(a) 12, 20, a', a

(b) 22, 16, b', b

(c) 6, 12, c', c

(d) 25, 6, d', d

Chapter 7

Descriptive Geometry

7.1 Introduction

Back in the 18th century 2011 French mathematician and engineer, Gaspard Monge (1746-1818), was involved with the design of military armoury. He developed a system, using two planes of projection at right angles to each other, for graphical description of solid objects.

This system is, called *descriptive geometry*, and provided a method of graphically describing objects accurately and unambiguously. It relied on the perpendicular projection of geometry from perpendicular planes. Monge's descriptive geometry forms the basis of what is now called *Orthographic and auxiliary projections*.

7.2 Orthographic projection

Orthographic projection is passing precise information about three- dimensional shapes or figures in a two-dimensional manner. Orthographic projection is a graphical method used in modern engineering drawing to provide accurate information about a given view of any object. In order to read, interpret and communicate with engineering drawings, a designer must have a sound understanding of its use and a clear vision of how the various projections are created.

Unlike other projections in which three views are observed at a time, orthographic projection view individual surfaces; top, bottom, front, back, and end views separately. These views are 'laid' out on paper in a systematic way so that when the views are 'read' the original solid can be visualized. These views are laid out on paper in a systematic way so that when the views are 'read' the original solid can be visualized.

The word *Orthographic* means to draw at right angles and is derived from the Greek words:

ORTHOS - straight, rectangular, upright

GRAPHOS - written, drawn

Orthographic projection is therefore a method of producing a number of separate 2D inter-related views, which are mutually at right angles to each other. Using this projection, even the most complex shape can be fully described. This method, however, does not create an immediate three-dimensional visual picture of the object, as pictorial projection.

Orthographic projection is based on projection on two principal planes — horizontal plane (HP) and vertical plane (VP) (elevation and end) — intersecting each other and forming right angles or quadrants. Take the solid below as example. Stand a piece of paper behind and look at it from the front as shown in Figure 7-1.

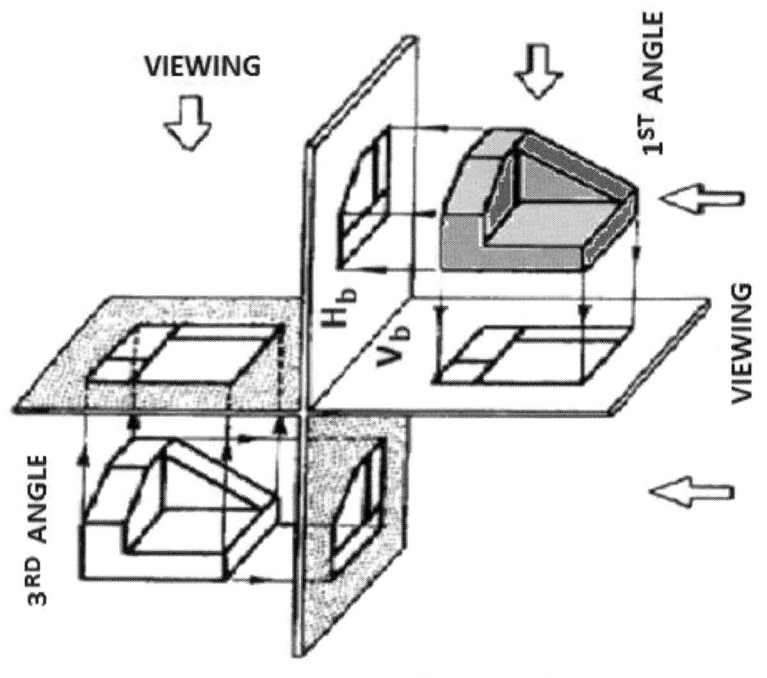

Figure 7-1: Views in different quadrants

Because the views are only two dimensional, more than one view is needed to completely describe the object. Usually two or three views is enough (front, top and side), but often more are required.

Traditionally, the views contain dimensions as follows:

    Front ⟹ Width and height/depth

    Top ⟹ Width and depth/length

    Side ⟹ Length and height

7.3 Multiview projection

To show views of a 3D object on a 2D piece of paper, it is necessary to unfold the planes such that they lie in the same plane with different views. You can also think of these views as an object inside a glass box (always observed from outside the box) with its surfaces "projecting" on to the sides of the box. You can then unfold the box to project the views on a flat surface. When this is done, you have the six views of the object.

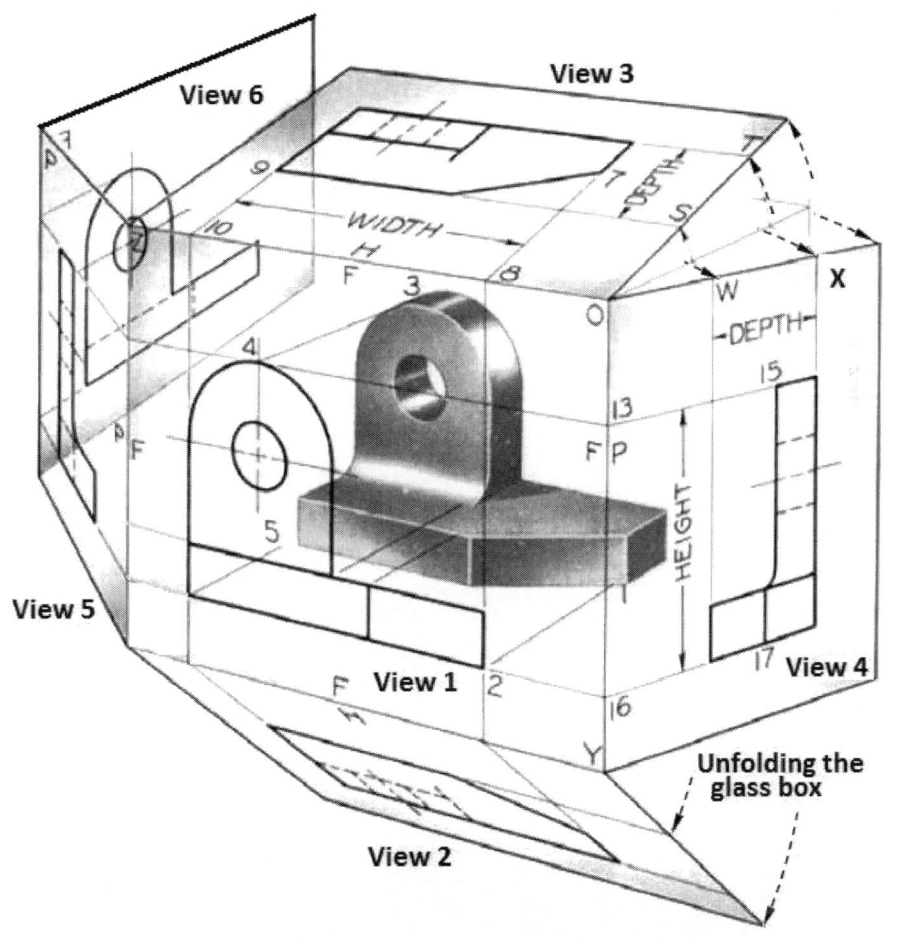

Figure 7-2: Object views in isometric box

All planes except the rear plane are hinged to the frontal plane, which is hinged to the left-side plane (Figure 7-3).

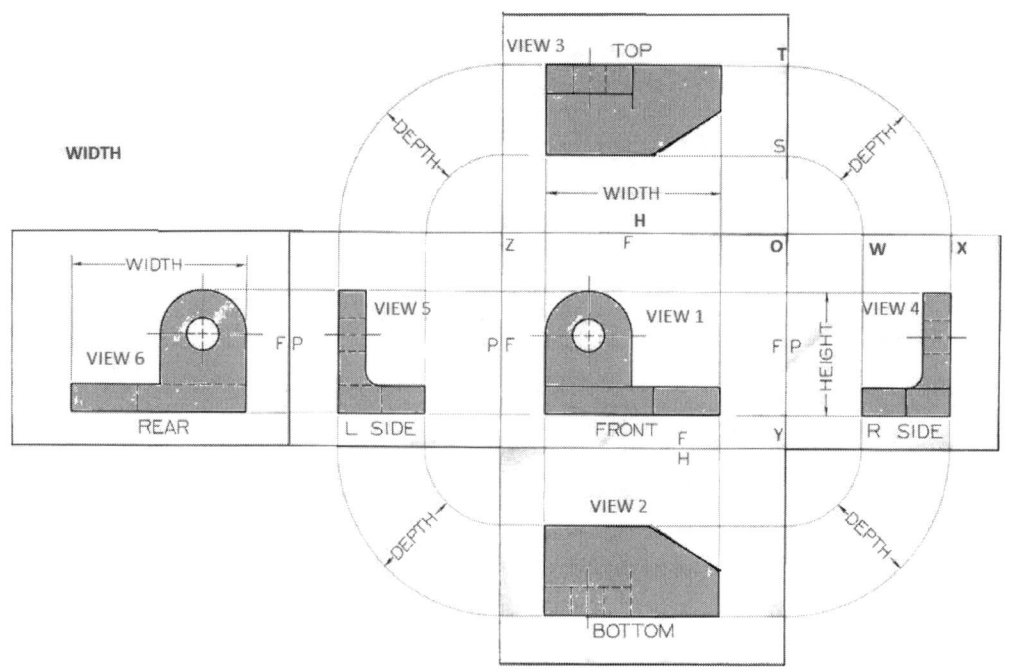

Figure 7-3: Object views in isometric box

It may not, be necessary to show all six views to completely describe the object. In fact, the minimum number of views is preferable. Imagine that you have an object suspended by transparent threads inside a glass box, as in (Figure 6-2). Then draw the object on each of three faces as seen from that direction.

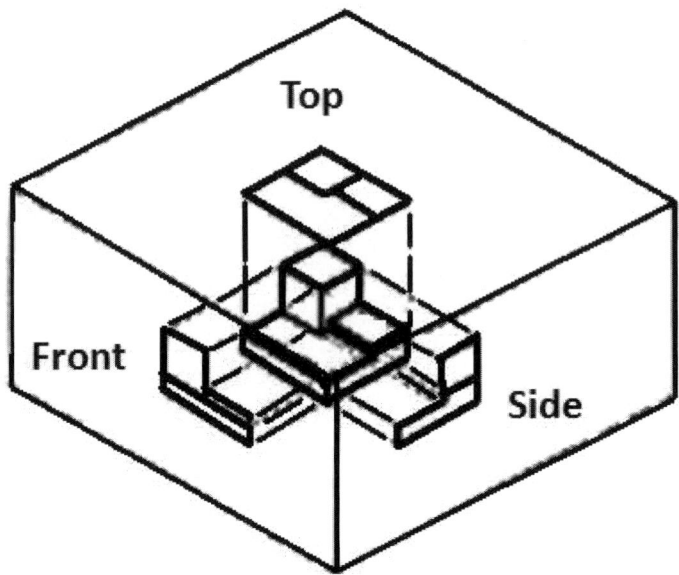

Figure 7-4: Object views in isometric box

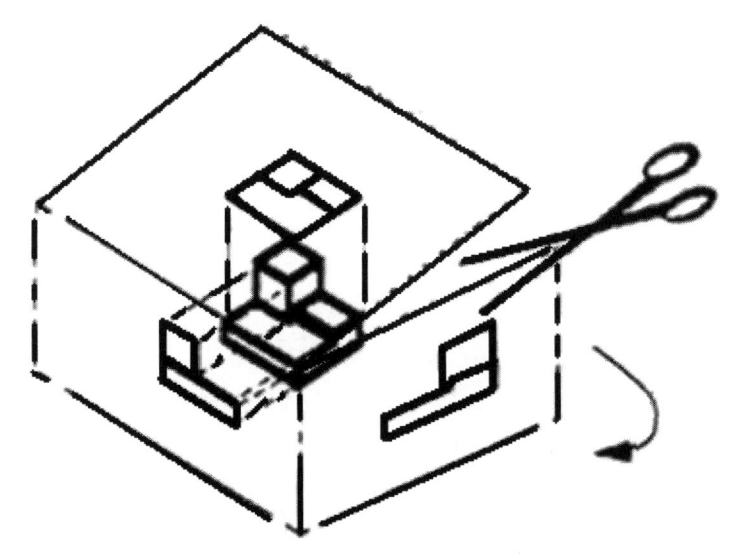

Figure 7-5: Cutting the isometric box

Figure 3: shows how the three views appear on a piece of paper after unfolding the box.

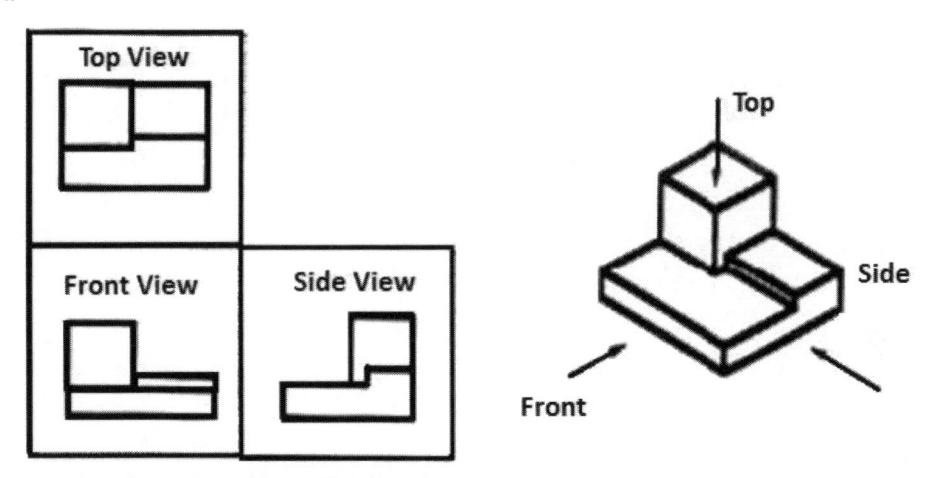

Figure 7-6: A Multiview drawing form the cut surface

Preferred view for a multiview drawing

The aim of an engineering drawing is to convey all the necessary information of how to make the part to the manufacturing department. For most parts, the information cannot be conveyed in a single view.

The views that reveal every detail about the object is often choose for a multiview drawing. Three views are not always necessary; we need only as many views as are required to describe the object fully. some objects need only two views, while others need four.

For example, how many views are necessary to completely describe this plate? 1? 2? 3? or 4?

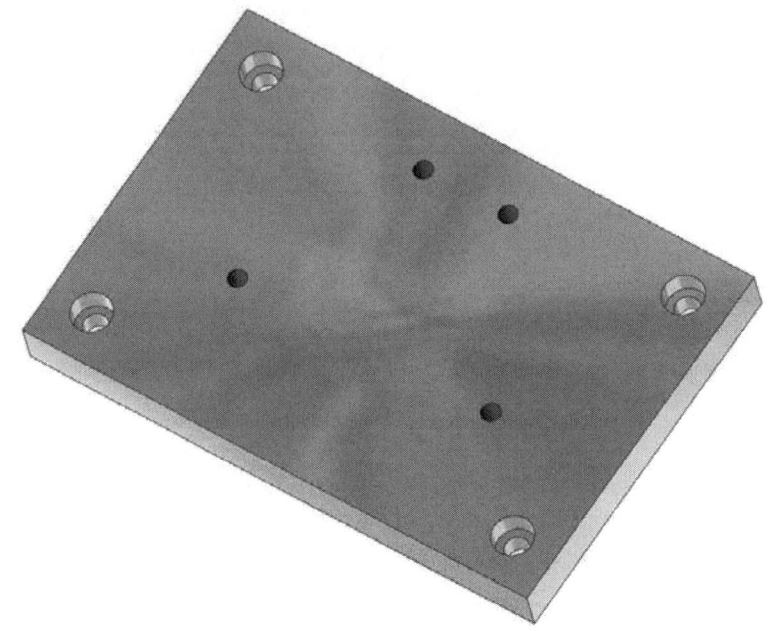

Figure:7-7: The answer is 2 views as shown below.

The answer is 2 views as shown below.

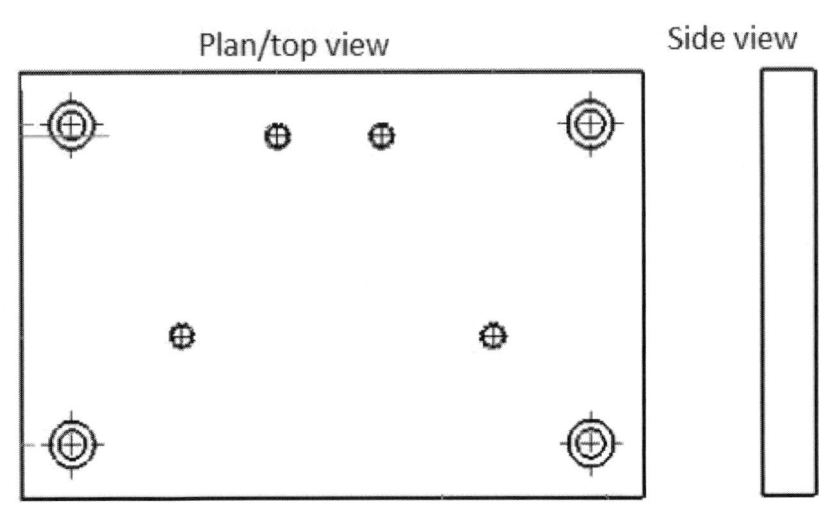

Figure 7-8: The answer is 2 views as shown below.

Preferred view for a circular object

The circular object such as cylinder, pipes, balls etc requires only two views (Figure 7-5).

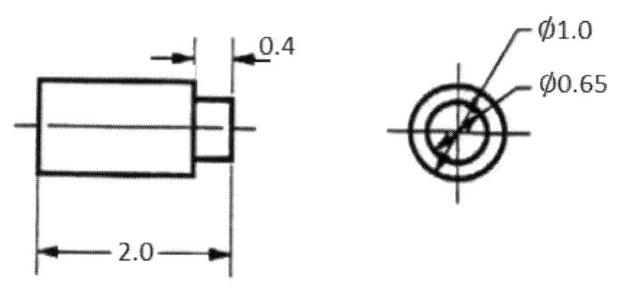

Figure 7-9: Types of multiview projection

There are two predominant *orthographic projections* used today. They are based on Monge's original right angle planes defined by four separate spaces, or quadrants. Space is divided by vertical and horizontal planes into four 90 angles objects can be placed in either 1st or the 3rd angle, as represented in Figure 7-10.

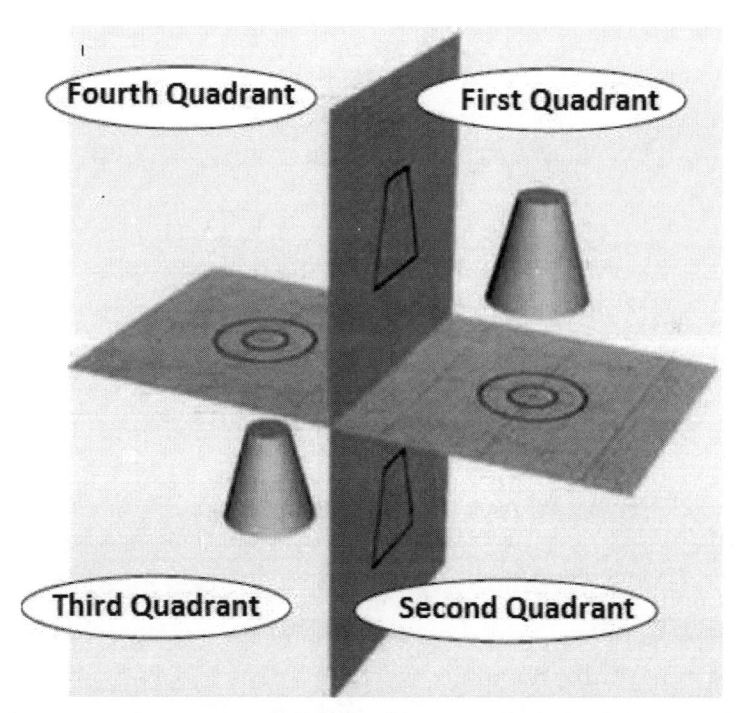

Figure 7-10: Projection in the quadrants

Each of these quadrants could contain the object to be represented. Traditionally however, only two are commonly used, the *first* and the *third*. The names First and

Third angles refers to the 1st and the 3rd quadrants of two interesting planes shown in Figure 6-6. If object is placed in the 2nd and 4th angles, the views will overlap when the planes are opened out.

Projections created with the object placed in the first quadrant are said to be in *First Angle* projection, and likewise, projections created with the object placed in the third quadrant are said to be in *Third Angle* projection.

Quadrant	Characteristics
1st	1. The vertical plane is a solid plane. Images registers on it 2. The horizontal plane is a solid plane. Images registered on it 3. Object placed between the planes and the source 4. Viewing in this place is possible
2nd	1. The vertical plane is a transparent plane. Images are reflected on it 2. The horizontal plane is a solid plane. Images registers on it 3. Object placed between the plane and the source, but the plan view is not easily seen from the viewing position above 4. Viewing in this plane is difficult
3rd	1. The vertical plane is a transparent plane. Images reflects on it 2. The horizontal plane is a transparent plane. Images reflects on it 3. Object placed between the plane and the source, but the plan view is not easily seen from the viewing position above 4. Viewing in this plane is possible
4th	5. The vertical plane is a solid plane. Images registers on it 6. The horizontal plane is a solid plane. Images registers on it 7. Object placed at the back of the plane and the source, but the object view is not easily seen from the viewing position above 8. Viewing in this plane is difficult

A simple demonstration of first and third angle projections can be arranged by placing the block on the drawing board and moving it in the direction of the four chain dotted lines terminating in arrowheads in Figure 7-11. Figure 7-11 shows the positioning for first angle and Figure 7-12 for third angle projection. The view in each case in the direction of the large arrow will give the five views already explained.

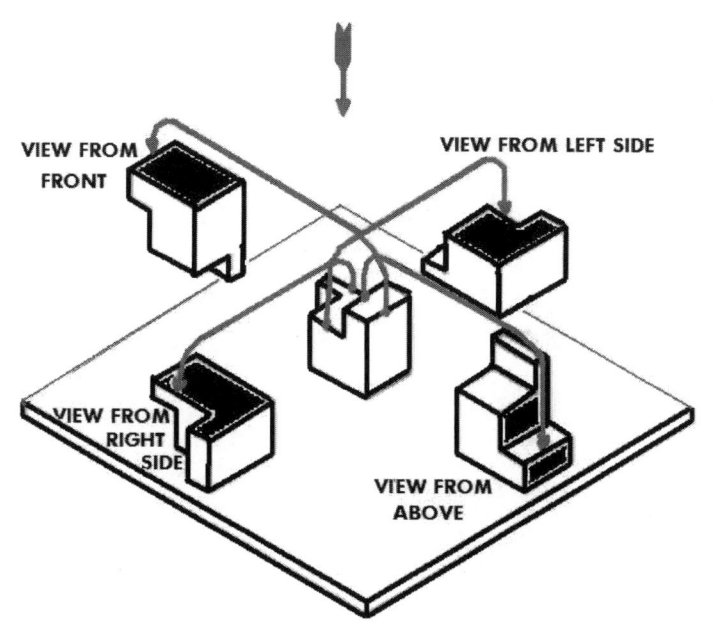

Figure 7-11: First angle arrangement

The terms; first and third angle, corresponds with the notation used in mathematics for the quadrants of a circle as shown in Figure 7-13. The block is shown pictorially in the first quadrant with three of the surfaces on which views are projected. The surfaces are known as planes and the principal view in direction of arrow A is projected on to the principal vertical plane.

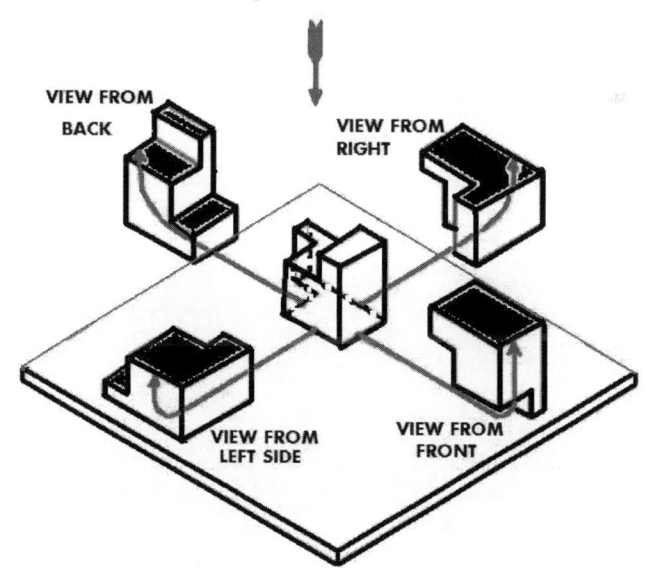

Figure 7-12: Third angle arrangement

The view from D is projected on to a horizontal plane. View B is also projected on to a vertical plane at 90° to the principal vertical plane and the horizontal plane and this is known as an auxiliary vertical plane. Another horizontal plane can be positioned above for the projection from arrow E, also a second auxiliary vertical plane on the left for the projection of view B. Notice that the projections to each of the planes are all parallel, meeting the planes at right angles and this a feature of orthographic projection.

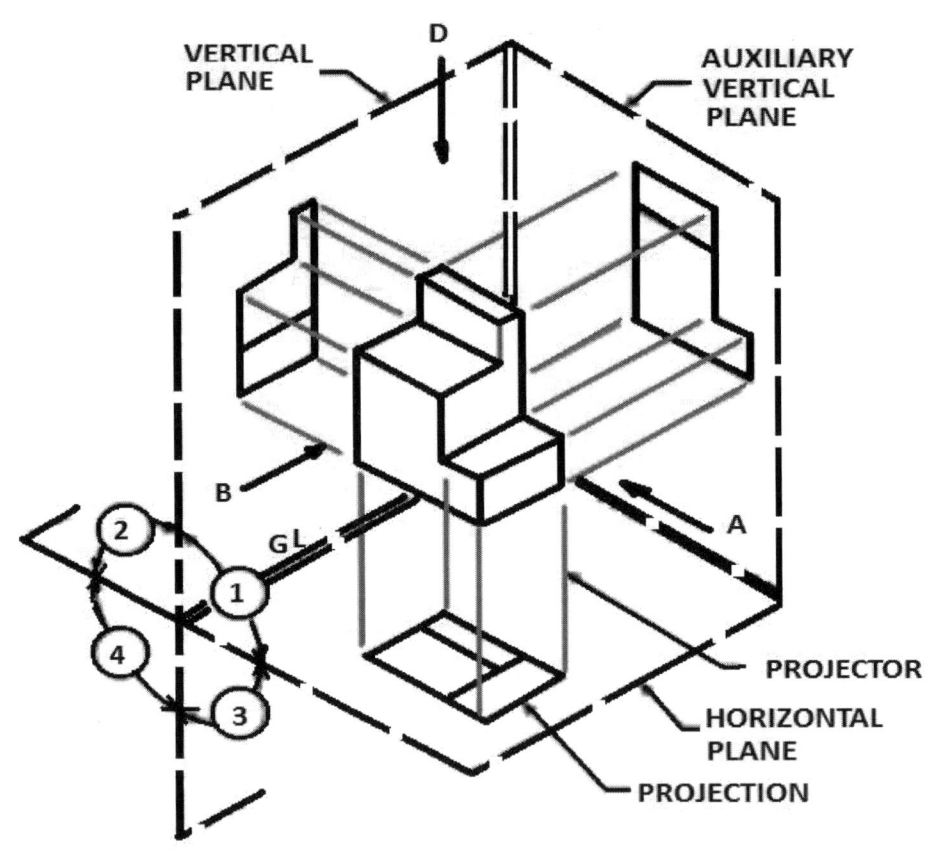

Figure 7-13: Projections in first quadrant

Notations
VP is the vertical plane.
HP is the horizontal plane.
AVP the auxiliary vertical plane
GL is the ground line.

The intersection of the vertical and horizontal planes gives a line which is the ground line GL. This line is often referred to as the XY line; this is useful in projection problems since it represents the position of the horizontal plane with reference to a front view and also the position of the vertical plane with reference to a plan view.

7.4 Principles of first angle orthographic projection

Consider the first quadrant in Figure 7-14A. The resultant drawing of the cone would be obtained by flattening the two perpendicular projection planes, as shown in Figure 6-10B. For this example, you could say that the right hand side image is the plan or *top elevation* and the image to the left is the *side elevation*. Traditionally, front views are also known as front elevations, side views are often known as side or end elevations and the views from above or beneath are referred to as plans.

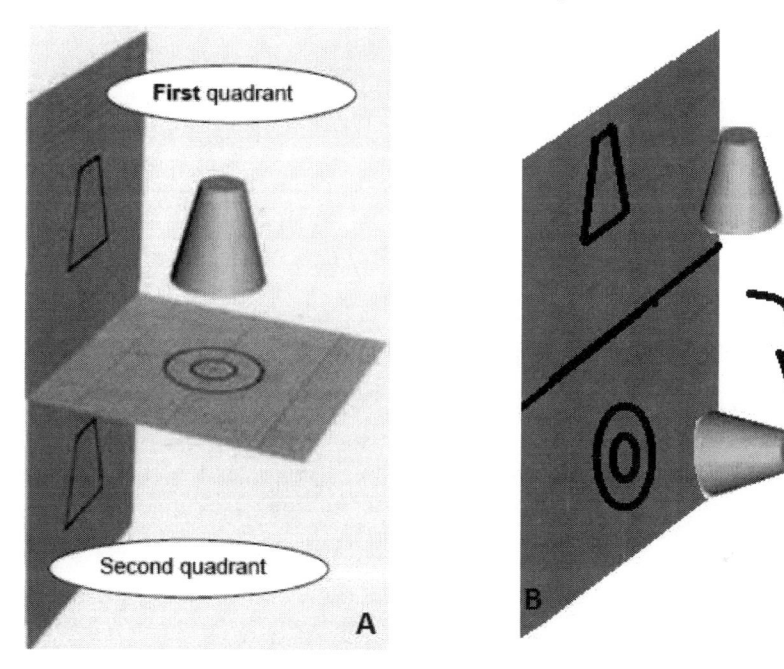

Figure 7-14: Views in first angle

Assume that a small block is made 35 mm x 30 mm x 20 mm and that two of the corners are cut away as shown below in three stages.

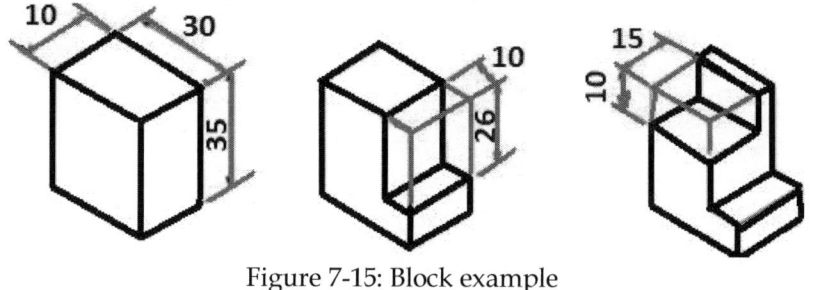

Figure 7-15: Block example

163 | Page

Figure 7-16 illustrates a pictorial view of the block and this has been arranged in an arbitrary way because none of the faces are more important than the others. In order to describe the orthographic views we need to select a principal view and in this case we have chosen the view in direction of arrow A to be the view from the front.

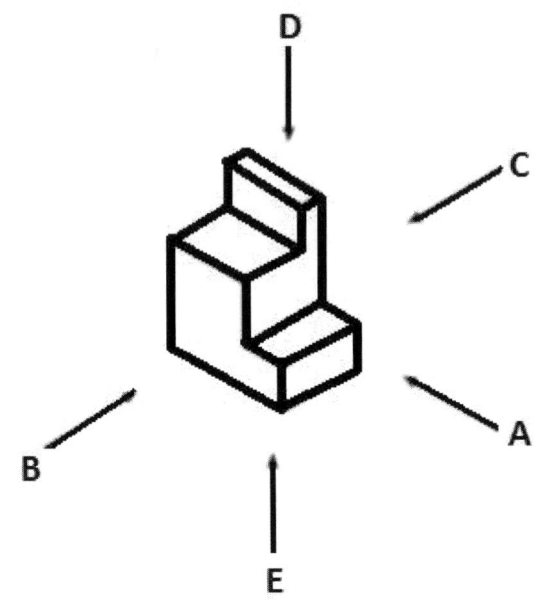

Figure 7-16: Illustrated block

The five arrows point to different surfaces of the block and five views will result. The arrows themselves are positioned square to the surfaces, that is at 90° to the surfaces and they are also at 90°, or multiples of 90° to each other. The views are designated as follows:

In Figure 7-16 the 1st angle projection views in the directions of arrows B, C, D and E are arranged with reference to the front view as follows:

View in direction A is the view from the front,

View in direction B is the view from the left,

View in direction C is the view from the right,

View in direction D is the view from above,

View in direction E is the view from below.

                    1ST ANGLE

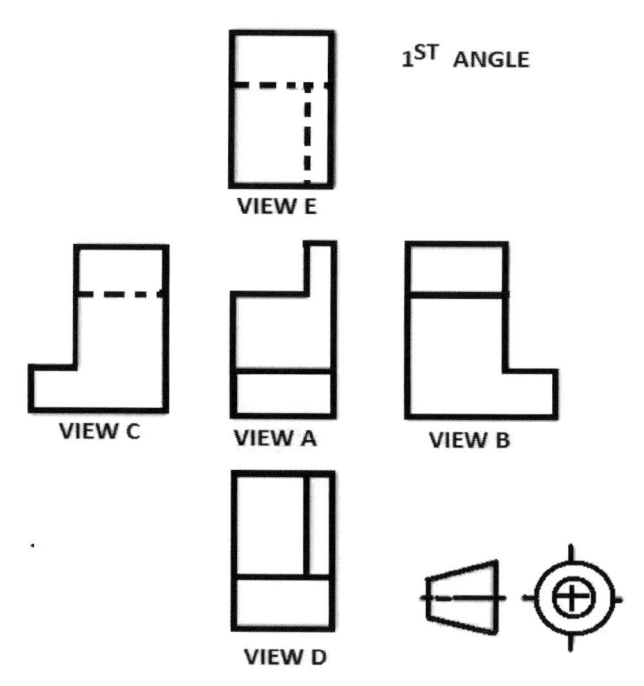

                    VIEW E

VIEW C   VIEW A   VIEW B

                    VIEW D

Figure 7-17: Views in first angle

The students are expected to know the above rules. It is customary to state the projection used on orthographic drawings to remove all doubt, or use the distinguishing symbol which is shown on the arrangement in Figure 7-18.

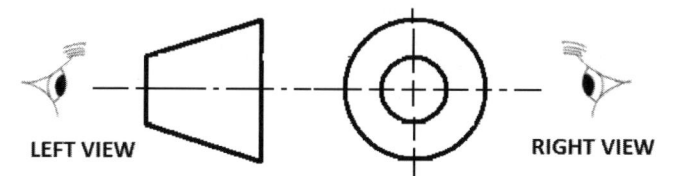

LEFT VIEW   RIGHT VIEW

Figure 7-18: First angle view

Whether you view the objects from the left or the right, the order in which the drawing views are arranged puts the image that you see *after* the *object*, i.e. *Object first, then the image*. This is always true for *first angle* view.

In another way, it can be put thus:

- *Viewing from the left*: The drawn image on the right is your view of the drawn object on the left.
- *Viewing from the right*: The drawn image on the left is your view of the drawn object on the right.

This can get confusing, particularly when also considering other drawings created using other projections. You may develop your own way of recognizing First Angle projection. The author uses:

The *Object* is *First* for *First* Angle view.

An example of a component represented in a multiview drawing, in first angle projection is shown below. The usual practice is to orient the component in a position that it is most likely to be found in.

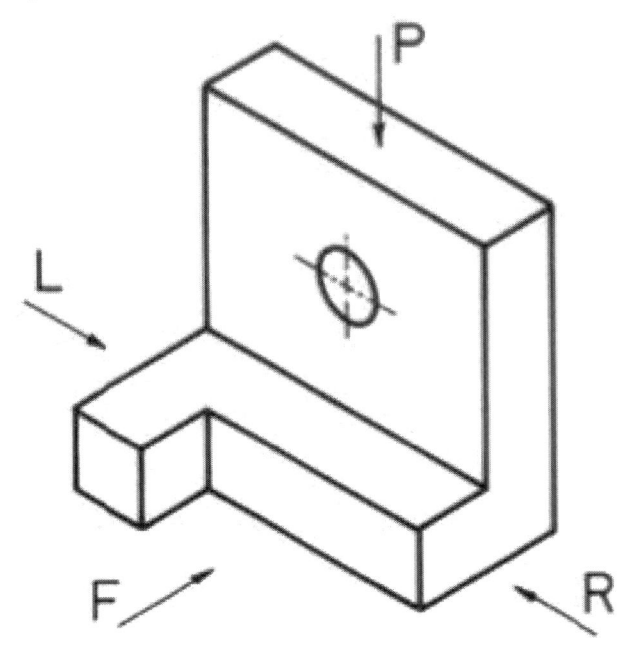

Figure 7-19: 1st angle projection line of views

Your aim is to create, from the front view, an orthographic projection drawing as shown below. Note how the views are constructed in line with each other, allowing the features to be 'projected' between the views. So, the stages are:

1. Choose which view direction or face will be used as the front view of the component.
2. Draw the outline of the front view, leaving room for the other views.
3. Draw feint construction lines out from the front view.

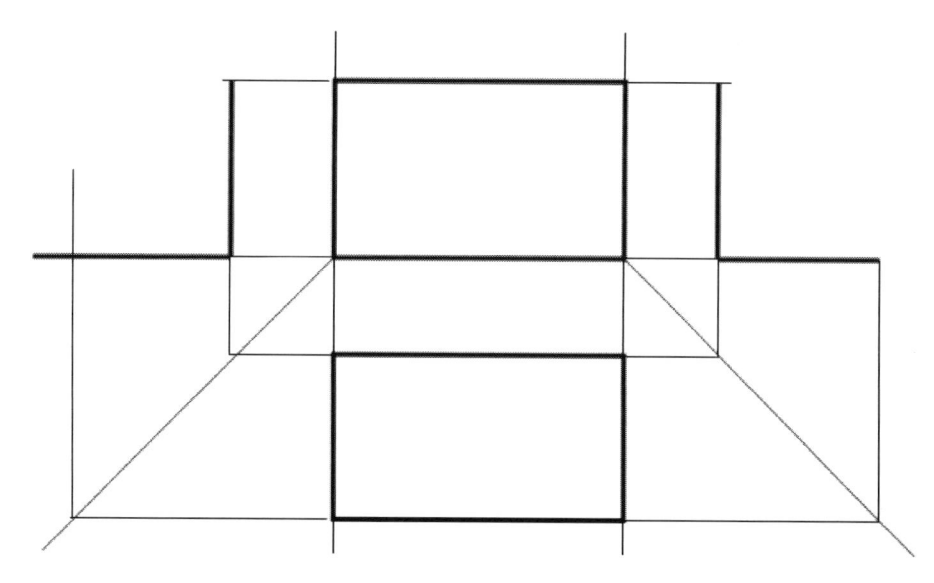

Figure 7-21: Construction line frame

4. Start to draw the outlines of the other views, using sides you know the length of.
5. Complete the details of the views by adding any required hidden detail lines, other outlines and center lines.

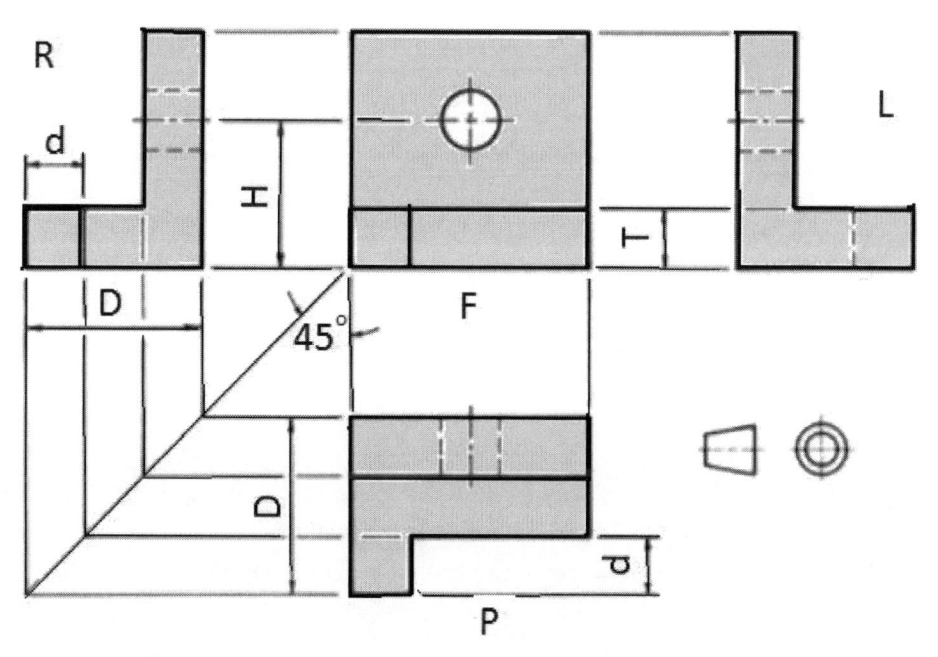

Figure 7-20: A completed first angle projection drawing

167 | Page

With first angle projection the plan view is *below* the front view. If you had placed the plan view *above* the front view it would actually have to become the bottom or underside view!

7.5 Principles of third angle orthographic projection

Third angle projection

The construction method used in 1st angle is the same for 3rd angle projections. For the same component, an orthographic projection drawing with the same front, side and plan views would look like as shown below.

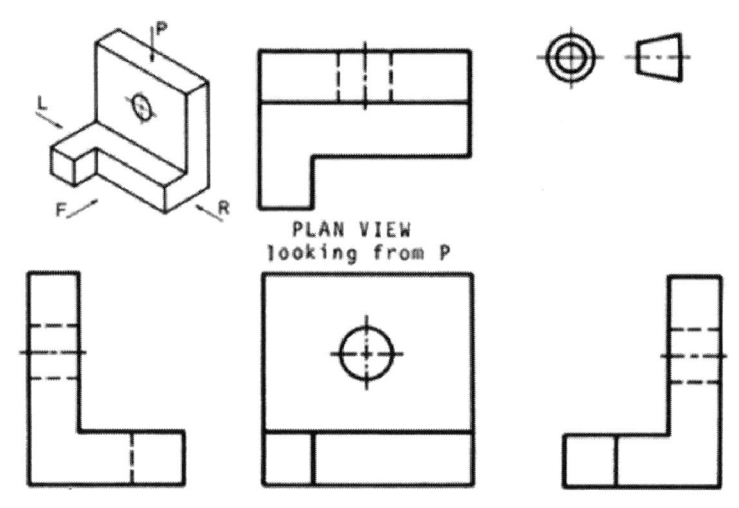

Figure 7-22: Third angle projections

Observe how, in third angle, the views give the image then the object. In other words, what you see then what you are looking at. Contrarily, in first angle you are given the object then the image, or what you are looking at, then what you see.

Third angle view

Consider the third quadrant in Figure 7-23A. The resultant drawing of the cone would be obtained by flattening the two perpendicular projection planes, as shown in Figure 7-23B. For this cone example, you would say that the left hand image is the plan or *top elevation* and the image to the right is the *side elevation*.

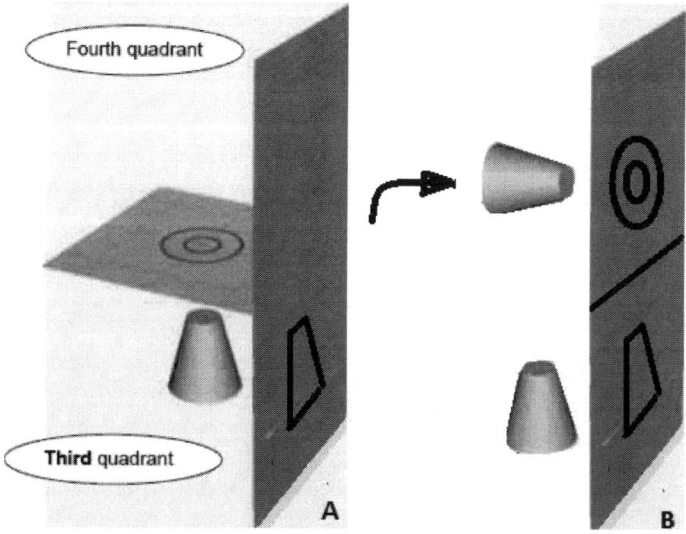

Figure 7-23: Third angle view

Considering the pictorial drawing illustrated in Figure 7-16 and replicated in Figure 7-24 below of the block, in order to describe the 3rd angle orthographic views we need to select a principal view and in this case we have chosen the view in direction of arrow A to be the view from the front.

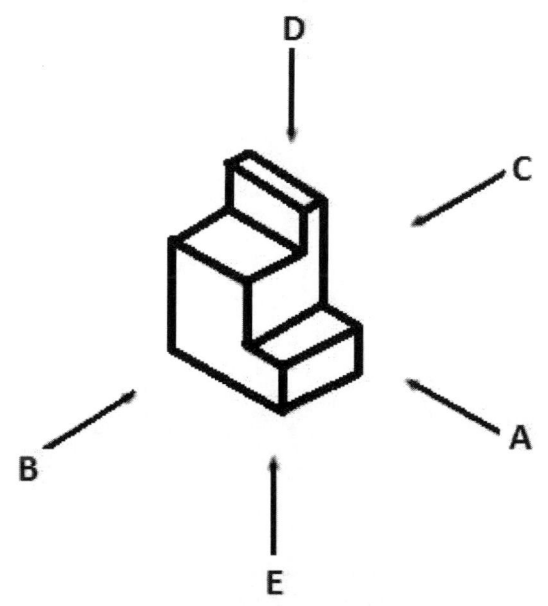

Figure 7-24: Illustrated block

The five arrows point to different surfaces of the block and five views will result. The views in the directions of arrows B, C, D and E are arranged with reference to the front view. The arrows themselves are positioned square to the surfaces, that is at 90° to the surfaces and they are also at 90°, or multiples of 90° to each other. The views are designated as follows:

The view from B is placed on the right,

The view from C is placed on the left,

The view from D is placed underneath,

The view from E is placed above.

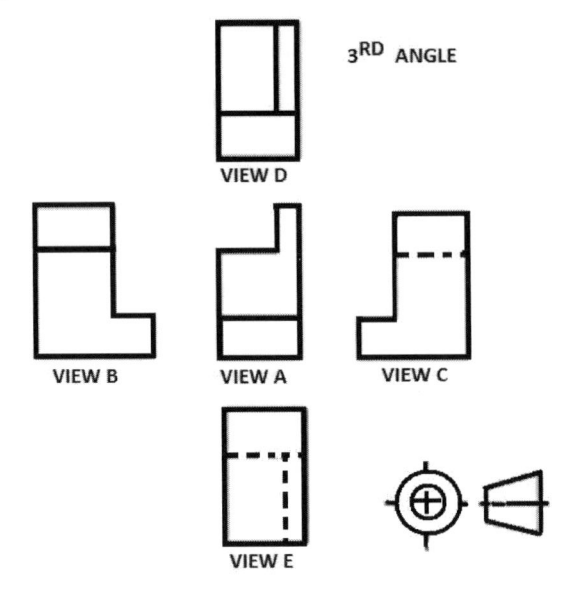

Figure 7-25: Views in third angle

Whether you view the objects from the left or the right, the order in which the drawing views are arranged puts the image that you see *before* the *object, image first then the object*. This is always true for *third angle* projection.

Put this in another way:

- *Viewing from the left*: The drawn image on the left is your view of the drawn object on the right.
- *Viewing from the right*: The drawn image on the right is your view of the drawn object on the left.

Again, you may develop your own way of recognizing third angle projection.

Perhaps: *Eye > Image> Object*

The same component is shown using *Third Angle* view.

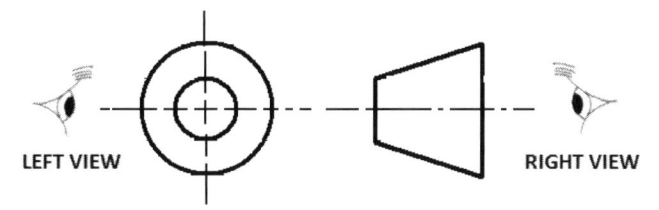

LEFT VIEW   RIGHT VIEW

Figure 7-26: Line of view in 3rd angle

7.6   Symbols for orthographic projection

Also there are conventions usually associated with views in first and third angles as show in the figures below.

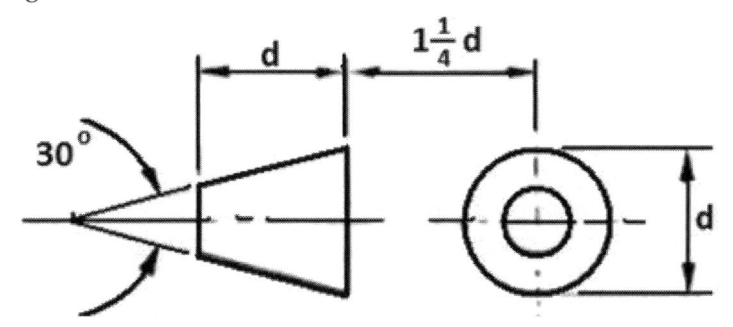

Figure 7-27: Recommended symbol proportions

Both systems of projection, first and third angle, are approved internationally and have equal status. The system used must be clearly indicated on every drawing, using the appropriate symbol shown in Figure 7-28 below.

Projection	Symbol
First Angle	
Third Angle	

First Angle projection is common in Europe

Third Angle projection is widely used in both the USA and the UK

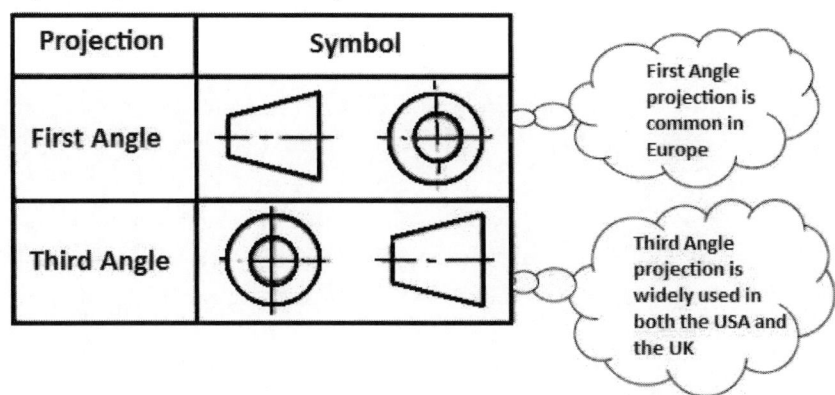

Figure 7-28: Projection system symbols recommended

First angle projection is widely used throughout all parts of Europe and often called European projection. Third angle in the system used in North America and alternatively described as American projection. In the British Isles, where industry works in co-operation with the rest of the world, both systems of projection are regularly in use. The current British and ISO standards states that both systems of projection are equally acceptable but they should never be mixed on the same drawing. The projection symbol must be added to the completed drawing to indicate which system has been used.

7.7 Comparing 1st and 3rd angle projection

S/No	Differences	1st Angle	3rd Angle
1.	Arrangement of views	The elevation view is above the plan view	The plan view is seen above the elevations
2.	Projection symbols		
3.	Adaptability	Widely used in Europe	Common in America
4.	Plane property	All planes are real and solid	All planes are transparent and virtual
5.	Object positioning	Object is positioned between the planes of projection and the line of view	The plane of projection is in-between the line of view and the object

6.5 Auxiliary projection

Auxiliary projection

Auxiliary projection is a view or projection of any line, surface or solid on a plane other than the principal planes of projection. Inclined planes and oblique (neither parallel nor perpendicular) lines appear foreshortened when projected to the principal planes of projection. Principal or ordinary planes include the Horizontal plane (H.P), vertical plane (V.P) and the second vertical plane (S.V.P). Any other plane other than the principal planes will be referred to as *auxiliary plane* (Figure 7-29).

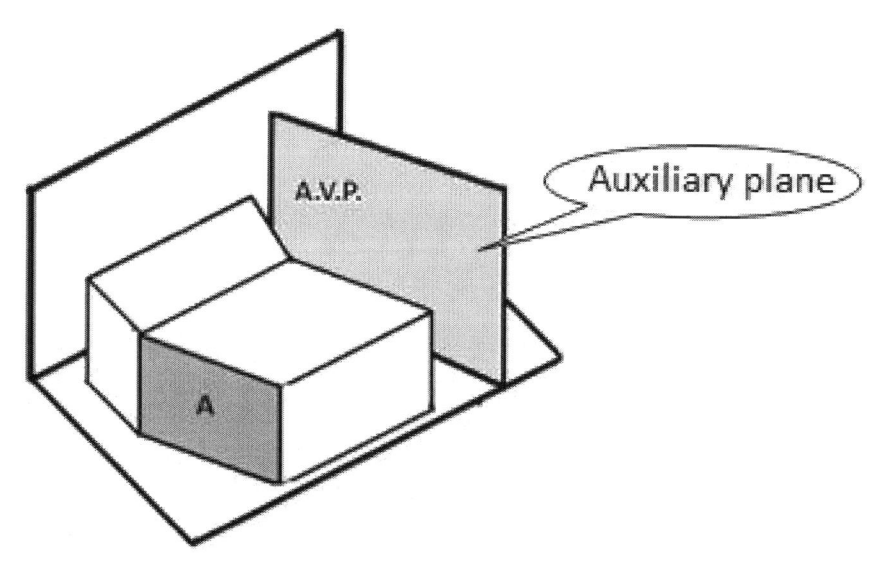

Figure 7-29: Position of auxiliary plane within principal planes

Uses of auxiliary views

1. Auxiliary views are used to find true length of an oblique line
2. To find the point of projection of a line
3. To find the edgewise view of a part
4. To find the true shape of a viewed surface

Drawing auxiliary views

To obtain a true size view, auxiliary views are created using similar techniques for creating standard views, unfolding about an axis. It should be noted that

a. The plan view shows only the true length of horizontal lines and
b. The elevation views show the true length of lines vertical or perpendicular to the horizontal and those which are parallel to the vertical planes.

But it is often necessary to obtain the true shape of a surface and the true length of a line before a material can be prepared for erection.

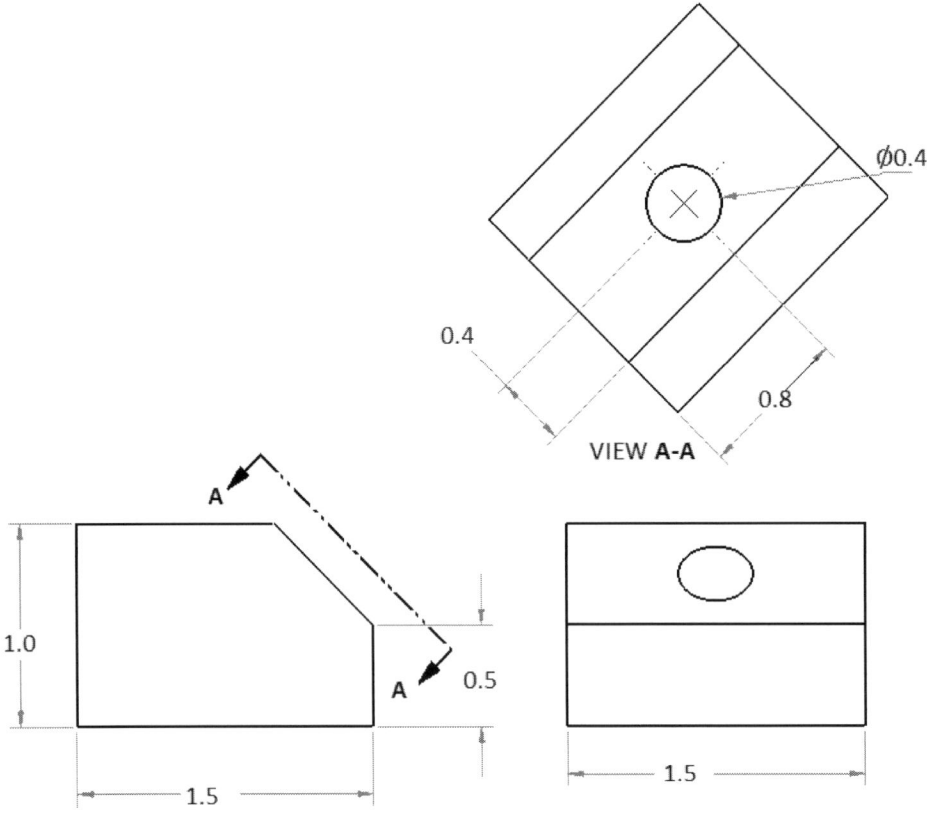

Figure 7-30: Auxiliary projection

Construction of auxiliary projections

Projections can be done in three views; auxiliary elevation auxiliary plan and combined projection. If for instance we are to draw the true view of an isometric inclined plane, we have to incline our position so as to be able to view the plane normally.

Two methods are generally employed in the construction of auxiliary projections:
1. Method of auxiliary projections (elevation or plan) and
2. Rotation method or rabatment.

Auxiliary elevation

To construct an auxiliary elevation, take your projections from the plan at 90° to the new XY plane produced and measure the lengths from the elevation.

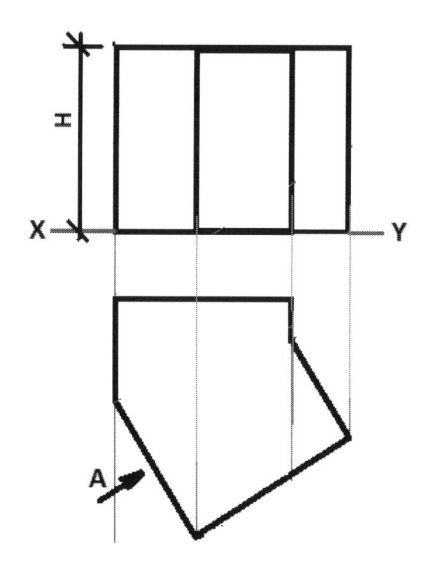

Figure 7-31: Plan and elevation view

For example, an isometric view of a block is shown in Figure 7-31 above. To show the true shape of the vertical face marked 'A', the block must be viewed at right angles to the surface, proceed as follows:

1. Draw the plan and elevation of the block as shown in Figure 7-32
2. Draw the new auxiliary plane X'Y' at 90° to the line of view A, indicated by the arrow.

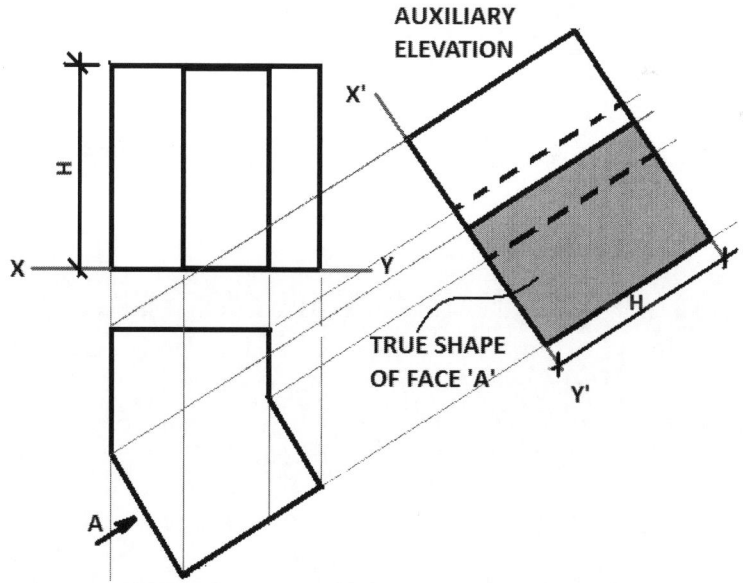

Figure 7-32: Drawing true shape in auxiliary elevation

3. Project the important points from the plan at 90° to the plane as shown
4. The heights taken from the front elevation are transferred to the new elevation termed the auxiliary elevation which shows the true shape of the surface.

Similarly, Figure 7-33 below shows a roof having a splayed end 'A' and the auxiliary elevation projected in similar manner to the previous example requiring the true shape of the end.

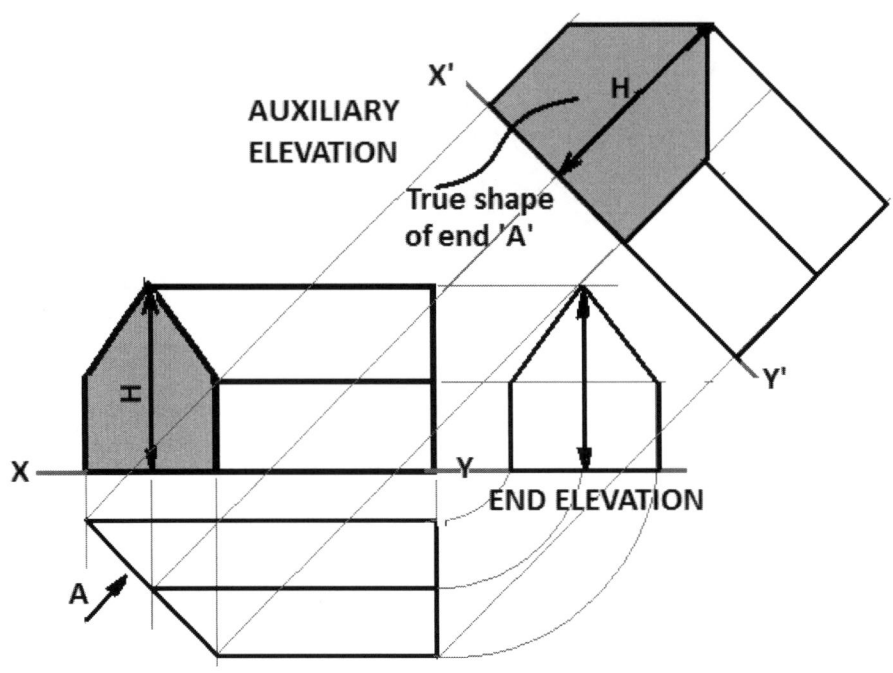

Figure 7-33: True shape in auxiliary elevation

Consider the isometric view of the truncated box in Figure 7-34. It is required to produce the true shape of the hatched surface in the direction of arrow.

The procedure is as follows.

1. Draw the front view and the plan views of the required surfaces
2. The projection of the inclined surface on the elevation is taken and the auxiliary plane drawn parallel to the line surface at the elevation.
3. Other dimension are taken from the plan to complete the true surface

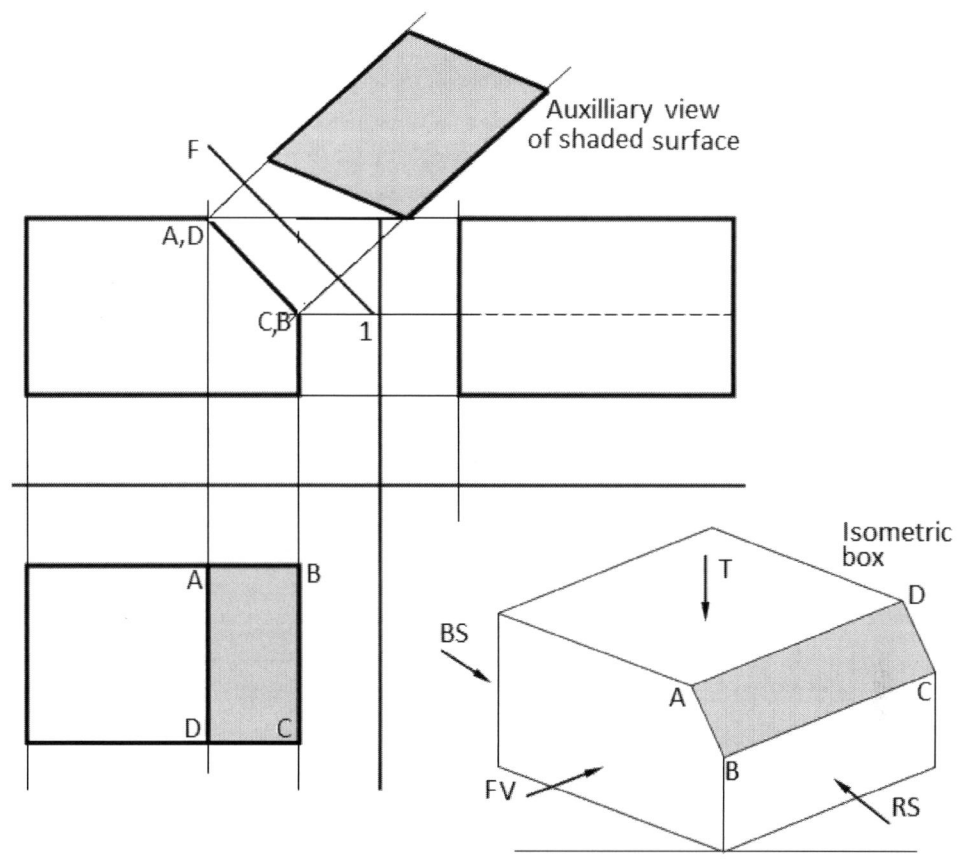

Figure 7-34: True shape of surface

Projecting auxiliary plans

Figure 7-34 below shows the plan and elevation of a square pyramid. If it is required to draw the auxiliary plan in a plane at right angles to the face $A'B'$, we can proceed as follows:

1. Draw the new ground line or $X'Y'$ line which must be at right angle to the line representing to face $A'B'$ in elevation.
2. Draw the new projections below $X'Y'$ with the corresponding distance to give the auxiliary plan as shown in Figure 7-35.

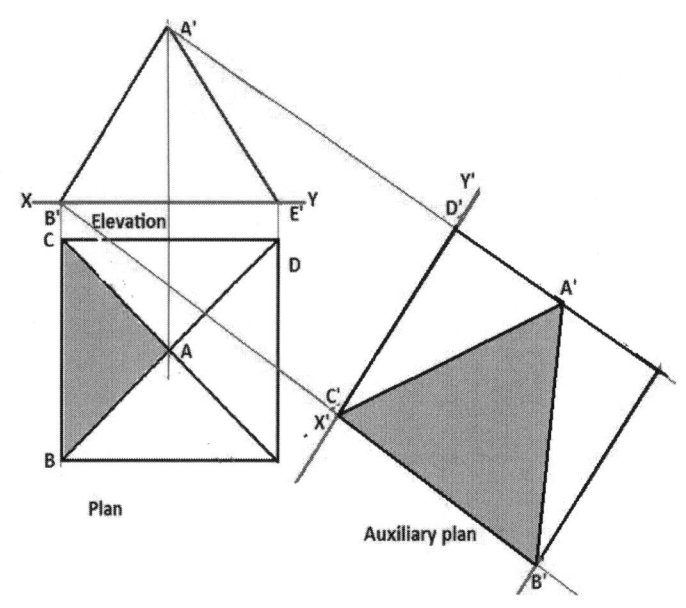

Figure 7-35: True shape in auxiliary plan

A further example is shown in Figure 7-36 below. The auxiliary plan is obtained by taking projections from elevation.

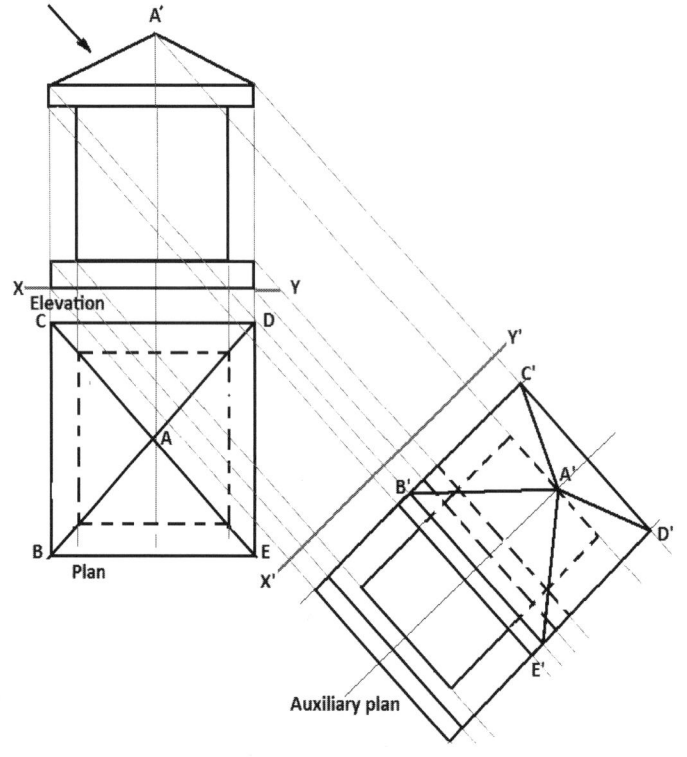

Figure 7-36: True shape in auxiliary plan

178 | P a g e

Combined projection

Apart from the auxiliary plan and elevations the third example of projections is combining the two elevations of projection as shown in Figure 7-36 below of an octagonal pyramid. In the figure above take the solid figure as example and stand a piece of paper behind and look at it from the front side and above, imagine the shapes being projected on the paper (plane). These three views can be brought together and stretches out in just one view as shown below.

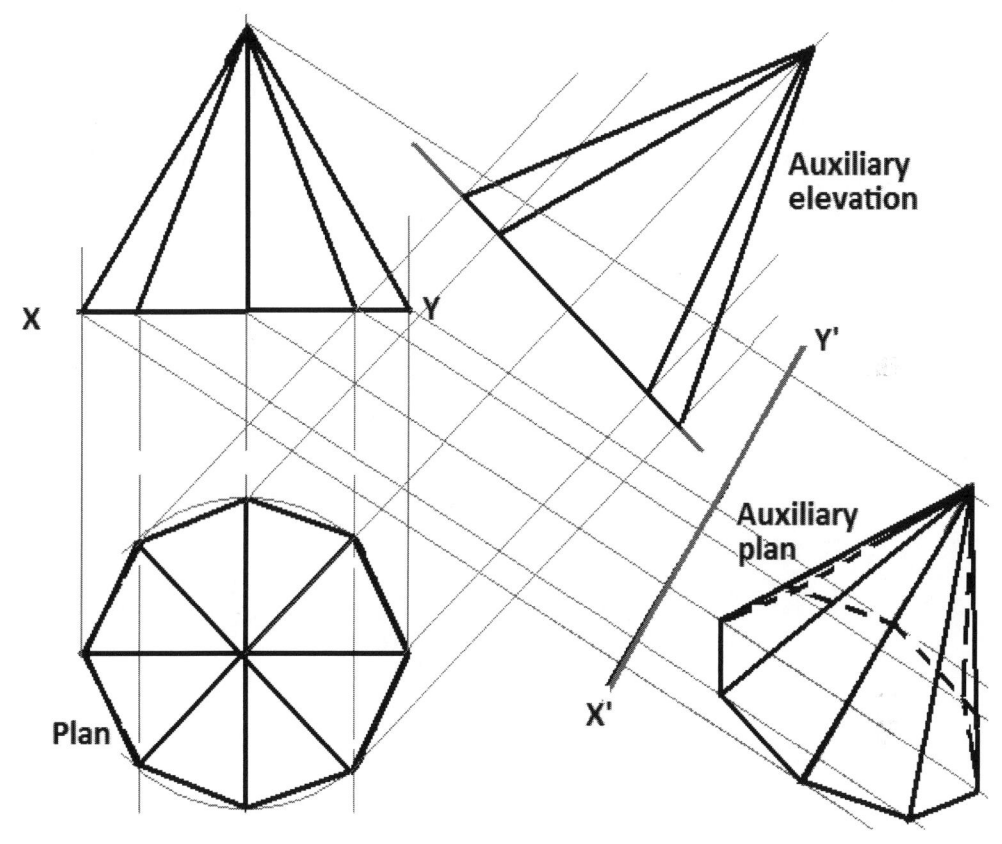

Figure 7-36: True shape in combined plane

Exercise

1. In the boxes below sketch the top view and the front view of each hole type.

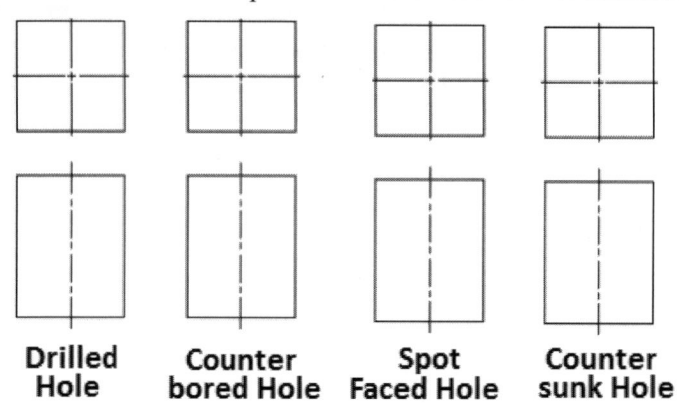

| Drilled Hole | Counter bored Hole | Spot Faced Hole | Counter sunk Hole |

2. Sketch the appropriate side view for each of the front views given below

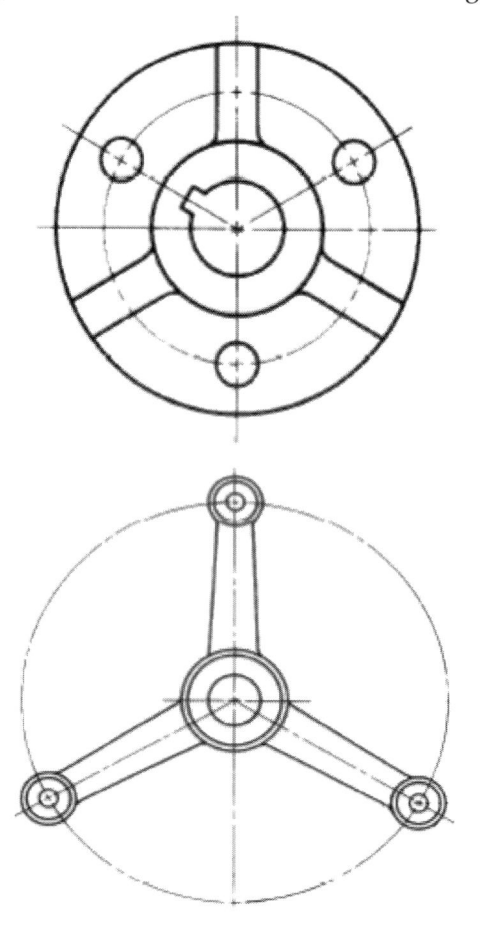

180 | Page

3. Find the missing lines in the orthographic drawings below

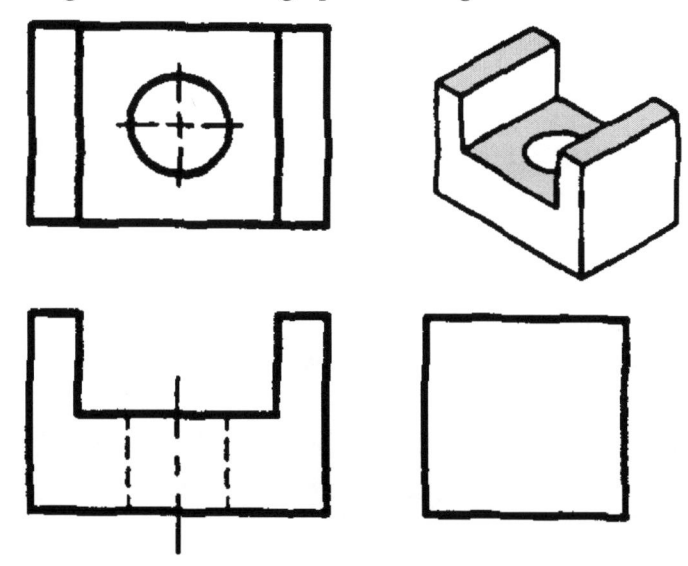

4. Find the missing lines in the orthographic drawings below

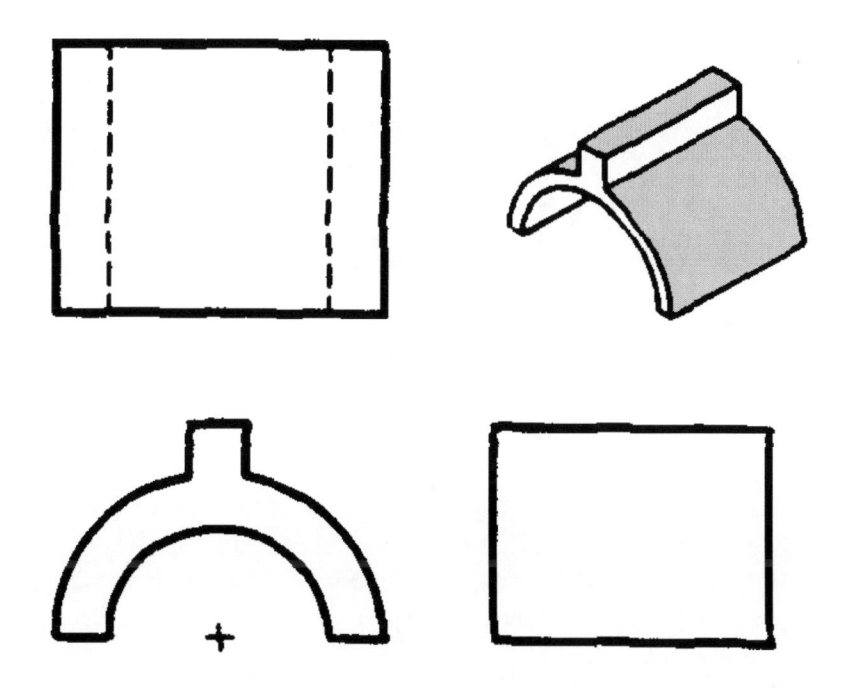

Chapter 8

Pictorial Drawing and Construction

8.1 Introduction

Pictorial drawings shows an object like you would see in a photograph. It give a three dimensional view of a room or structure. An example of pictorial sketch of kitchen is as shown in the Figure 8-1 below.

Figure 8-1: Pictorial drawing of a kitchen

Some of the most common difficulties in comprehension of illustrations involve close-up illustrations where a part, e.g., a person's hands or head, is used to represent the whole. While too much detail, particularly in the background may be confusing, outlined or stick figures contain too little detail, and are not as recognizable as toned-in line drawings. Perspectives, where objects in the distance are drawn smaller can present difficulties in the same way the pictures of small items can, e.g., insects, drawn to a much larger size than the actual sizes.

8.2 Projection of pictorial drawing

Projections are often useful in presenting a proposed object e.g. building to someone who is not familiar with the presentation in plan, section and elevation drawings. However, the rural population, in particular illiterates, may understand pictures and illustrations in a different way than intended or not at all. Even the idea that a message can be contained in a picture and that something can be learned from it can be new. This is mainly because they do not see many pictures and have not learned to understand the symbolic language often used in illustrations.

Types of pictorial drawing

The following types of pictorial projection are in use:

- Isometric
- Oblique
- Axonometric
- Perspective

Other types of pictorial drawing include

1. Diametric drawing
2. Assembly drawing
3. Sectioning
4. Free hand sketching

8.3 Isometric drawing

Isometric projection is a system of the representation of pictorial objects in three dimensional views. The representation shown in Figure 8-2 is called an isometric drawing. This is one of a family of three-dimensional views called pictorial drawings.

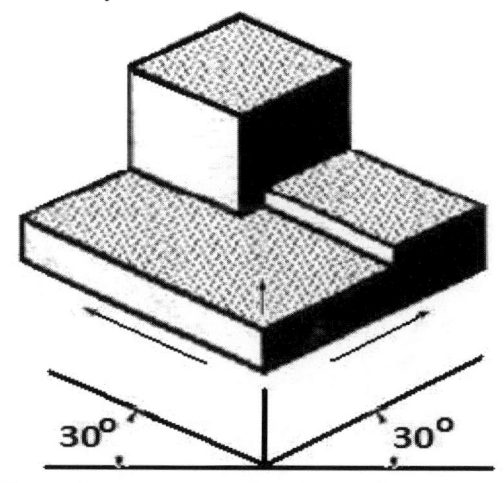

Figure 8-2: An isometric of a machined block

In an isometric drawing, the object's vertical lines are drawn vertically, and the horizontal lines in the width and depth planes are shown at 30 degrees to the horizontal. When drawn under these guidelines, the lines parallel to these three axes are at their true (scale) lengths. Lines that are not parallel to these axes will not be of their true length.

In isometric drawings, all dimensions along all the three axes are drawn to TRUE SIZE. Isometric projection is preferred when the three views of the object are of equal importance for accurate presentation of the object. Example: Cube of length L (Figure 8-3)

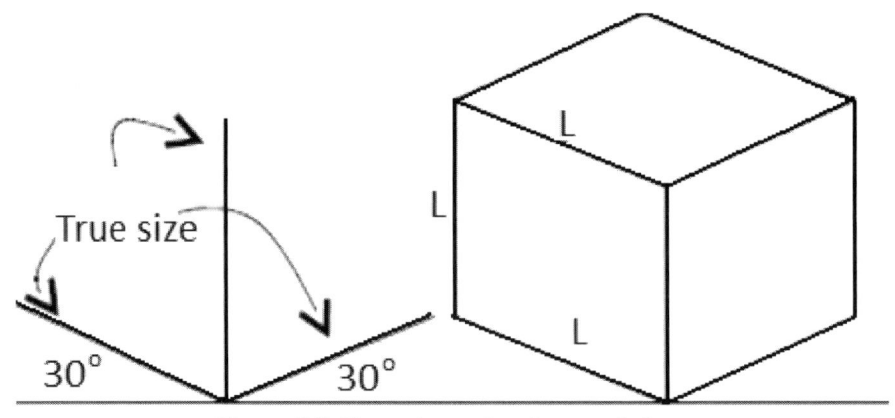

Figure 8-3: True sizes of an isometric box

Views in isometric drawing

There are three faces or views of an object of rectangular form can be shown and seen in one isometric drawing but not more than that. Since there are faces, two drawings will show them all, to do so effectively, the views must be selected to the best advantage.

Isometric box (cube)

A rectangular block is shown with its edges in the direction of the isometric axes. Now imagine them to represent the directions of the three edges at the corner of a rectangular box (Figure 8-4), the other views of the box can be completed by drawing through the extreme points we have marked.

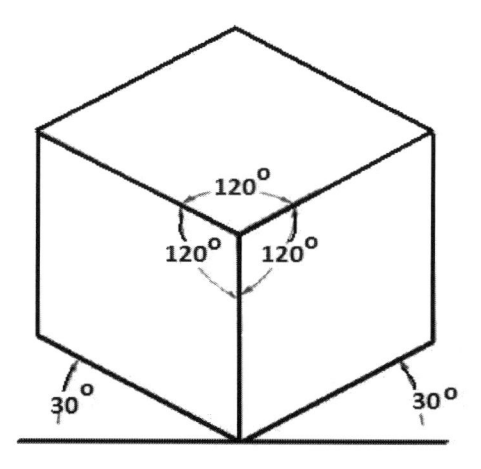

Figure 8-4: An isometric box

Note all lines are either:

- Inclined at an angle of 30° or
- Vertical except in oblique views.

Any engineering drawing should show every detail such that a complete understanding of the object should be possible from the drawing. If the isometric drawing can show all details and all dimensions on one drawing, it is ideal. However, if the object in Figure 7-2 had a hole on the back side, it would not be visible using a single isometric drawing. In order to get a more complete view of the object, an orthographic projection may be used.

Construction of isometric circle

It must generally be understood that all circles in isometric view will appear as an ellipse and vice versa. The isometric circle is drawn employing one of the following methods.

Method 1

Isometric circle may be constructed in a square of the same dimension as the radius of the circle. There is a simple and quicker method of constructing isometric circles using compasses and four points for circular arcs. The step by step method is escribed and illustrated in Figure 8-5 below.

1. Draw the isometric box and select the plane which will contain the isometric circle.

2. Draw the square and the inscribed circle radius R.

3. The selected plane ABCD forms an isometric square having two centrelines EF and GH drawn parallel to the sides as shown in Figure below.

4. Join points E to B and D to F. Also join points D to G and point H to B.

5. Mark the points of intersections as I and J to obtain centers for the small arcs radius r. Let points B and D be the other centers and radius R as the radius of the other arcs to complete the ellipse.

6. With centers B, D, I and J, and radii R and r in the figure, describe the arcs to complete the isometric circle.

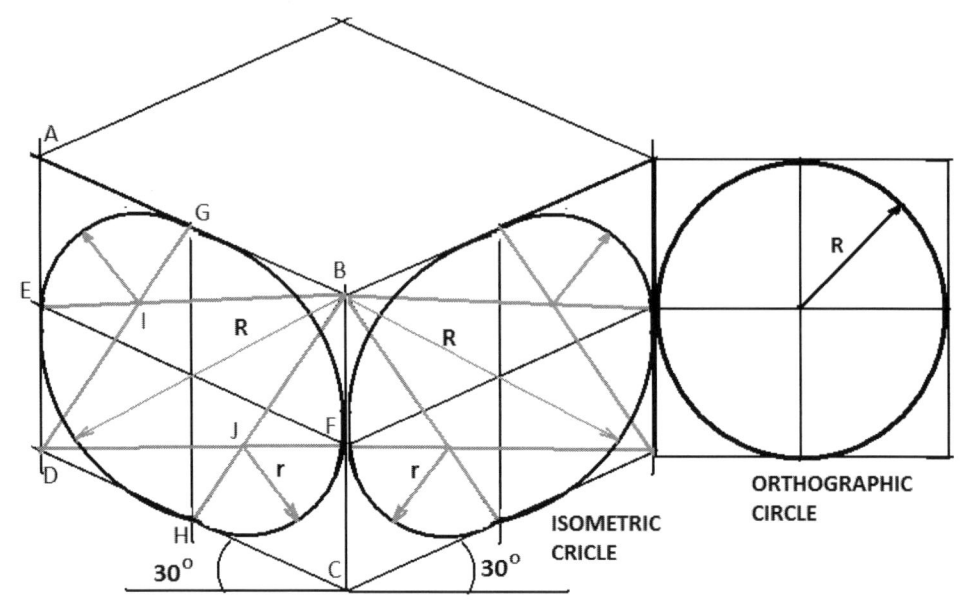

Figure 8-5: Square method of isometric circle

Same step can be repeated to draw isometric circles on any of the faces of the box.

Method 2: Drawing approximate isometric circle

Three stages are involved in constructing an approximate isometric circle as follows:

1. Draw the square with two center lines drawn parallel to the sides
2. Draw two horizontal lines and two inclined lines
3. At the intersection of the lines, mark the points 1 and 2

i. With points 1 and 2 as centers, draw arcs as shown
ii. With points 3 and 4 as centers, draw the closing arcs to complete the figure.

Method 3

Another method of construction is plotting many points (eight key points are usually enough) to enable us to draw a fair and accurate curve through them. The resulting curve is an ellipse. The longest diameter of the ellipse is called its major axis and the shortest diameter called minor axis.

The procedure for drawing this type of circle is highlighted below

1. The isometric square which will contain the isometric circle is first drawn
2. From one edge of this square draw a true square with its inscribed circle.
3. A number of points on the true circle are chosen and these are transferred to the isometric view.
4. A point is shown to illustrate the method of transferring any point.

An example of such procedure is described below

1. Draw the circle of a given radius on the center lines AA and BB. Complete the square as shown in orthographic view shown below
2. Divide the center line AA into equal parts and erect perpendiculars or ordinates as 1, 2, 3, 4, 5, 6, 7, 8,
3. Draw A'A' at 30° and B'B' vertically as center lines, equal AA and BB.

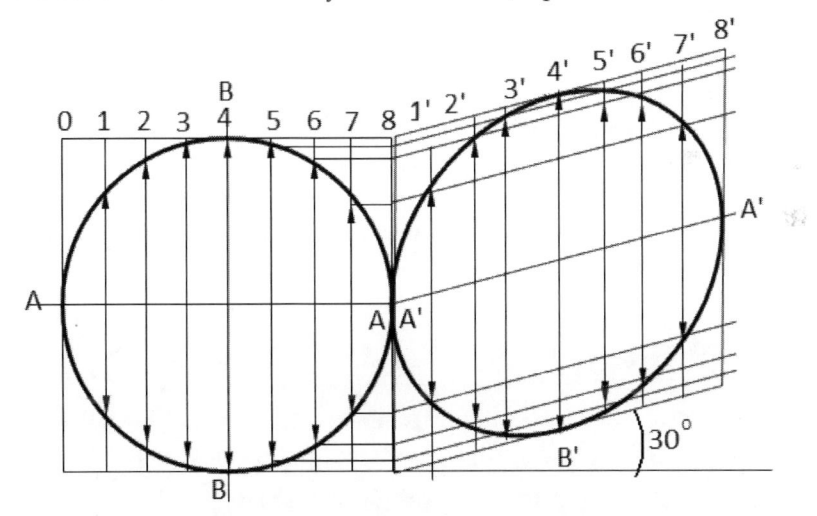

Figure 8-6: Inscribed circle method

4. On A'A', step off ordinates as on AA.
5. From A'A', step off on either side widths where the circle cuts each ordinate on 1',2'...7' as shown to give the isometric circle which is an ellipse.

Producing isometric drawings using Mat plans

A mat plan is a top view of a solid, with the number of cubes appearing in each vertical column displayed in the corresponding box. When given a mat plan, students may explore a drawing by building a three dimensional figure from such given plan using cubes. For example, the mat plan of set cubes in isometric view is shown in Figure 8-7. The following image is projected in three dimensions:

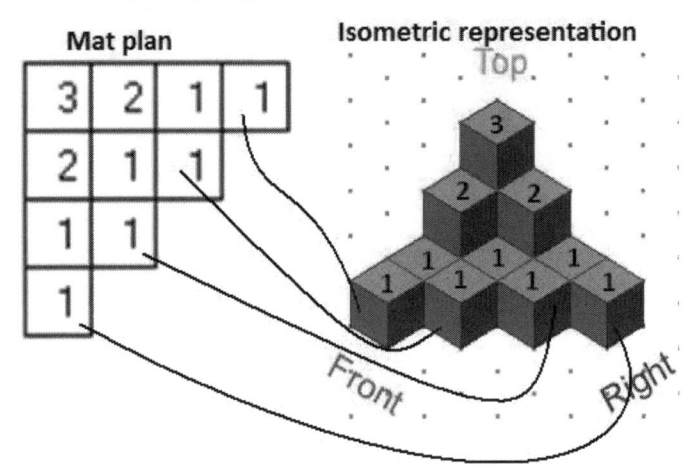

Figure 8-7: Isometric representation of mat plan

8.4 Oblique drawing

An oblique drawing is formed when front view projection of an object starts on the horizontal lines while the adjacent sides are then drawn to an angle, usually 30° or 45°, from the horizontal. An oblique projection starts with a front view of the building. The Views can be at any angle; 15, 30 or 45 degrees are common. The horizontal lines in the adjacent side are then draw to an angle, usually 30° or 45°, from the horizontal. The dimensions on the adjacent side are made equal to 0.8 of the full size if 30° is used or 0.5 if 45° is used. Curved and slanted lines are constructed in the same manner as in isometric projections.

The front view of an oblique drawing is drawn like it would be using orthographic projection. The front view shows all features with true shape and size while the top and side view are then projected back from the front view.

Two types of oblique drawings

Two types of oblique drawing are identified namely; cavalier and cabinet drawings.

Cabinet oblique

In cabinet drawings, the front face of the object is generally more important and complicated than the sides, and the depth is usually much less than the length or height. An example of cabinet oblique drawing is shown in the figure below. Different scales may be used in such drawing. This drawing is often used to draw furniture cabinets.

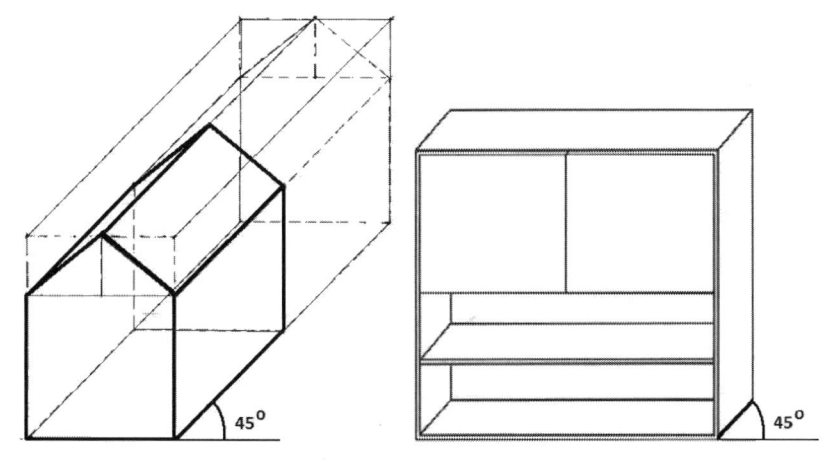

Figure 8-8: Cabinet oblique drawings

Cavalier oblique

In cavalier drawings the entire drawing uses the same scale in which all needed dimensions are scaled directly from the drawing. They sometimes create distorted appearances. Considering the drawing in Figure 8-9, although it is a true projection of the figure, but the side appears too long. This is caused by lack of providing allowance for the perspective or apparent convergence of the receding (diminishing) parallel lines.

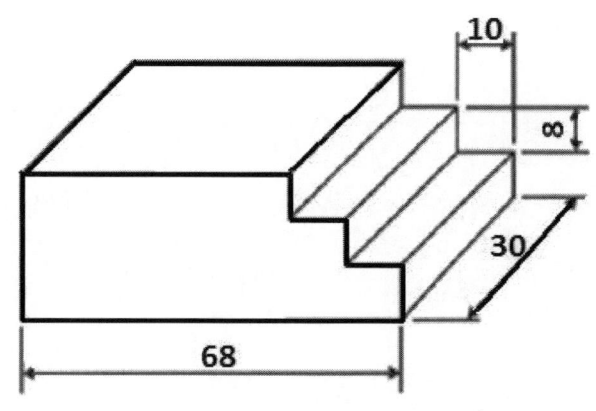

Figure 8-9: Cavalier drawing

189 | Page

Scale of the receding axis of such drawing will apparently have no effect on the true nature of the projection since the scale may be changed by altering the angle which the projectors make with the projection plane. If full size is used on the receding axis, the view is said to be in cavalier projection.

8.5 Axonometric drawing

These are drawings in which the object is drawn in three dimensions (3-D), i.e. three sides of the object appear in one drawing. Normally only one drawing is prepared or used.

- They are used extensively in artistic drawing.
- A three dimensional view (i.e. shows length, width and height of the object simultaneously)
- Provides only a general impression of the shape of the object by allowing the observer to see three of its sides as well as its three overall dimensions
- An exact and complete description of its shape, particularly as applied to its slots on the underside is lacking.

In axonometric projection the plan view of the building is placed on the drawing table with its side inclined from the horizontal at any angle. Usually 30°, 45° or 60° is chosen since those are the angles of the set squares. All vertical lines of the building remain vertical and are drawn to the scale of the plan view.

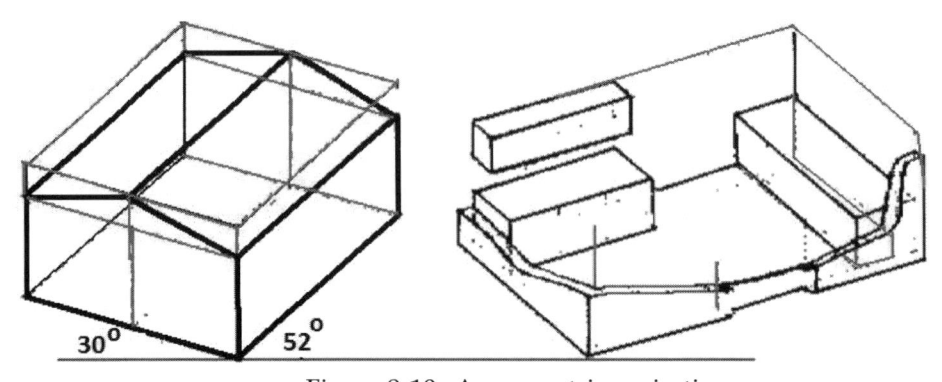

Figure 8-10: Axonometric projection

Two standards are currently used for axonometric projections: diametric projection and isometric projection.

8.6 Perspective drawing

The different technical terms used in perspective drawing can be explained if you imagine yourself standing in front of a window looking out at a building at some angle so that two sides of the building are visible. Trace on the window pane the outline of the building as you see it through the glass. You have then just made a perspective drawing of the building and if the glass could' be removed and laid on the drafting table the drawing would look like any perspective drawing made on paper.

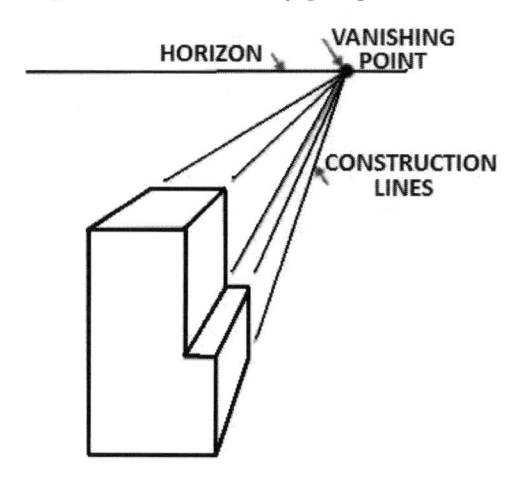

Figure 8-11: A typical parallel perspective drawing

Station point is the viewing point, supposedly occupied by the eye of the observer. The viewing point is also determined by the eye level, usually assumed to be 1.7m above ground level. Looking across a large body of water or a plain, the sky and water/ground appears to meet in the distance - the horizon line. This must always be considered present even when hidden by intervening objects.

The horizon line is at eye level. When looking down a straight road, the edges of the road appear to meet at a point (*the vanishing point*), on the horizon line. Similarly parallel horizontal lines of a building appear to meet at vanishing points, one for each visual side. The outline of the building was brought to the window by your vision - vision rays. The picture was traced on the window pane, which therefore can be called picture plane.

Two types of perspective drawing

1. Parallel (one-point) perspective:

- One face of the object is shown as the front view

- Lines parallel to the front view remain parallel
- Lines that are perpendicular to the front view converge at a SINGLE VANISHING POINT

2. Angular (two point) perspective: there are two categories;

- Parallel Perspective (One Point)
- Angular Perspective (Two-Point):
  o Similar to isometric drawings
  o One edge of the object is place in front
  o The two faces that meet at this edge recede to *different vanishing points*

All lines parallel to each face go to the different vanishing Points

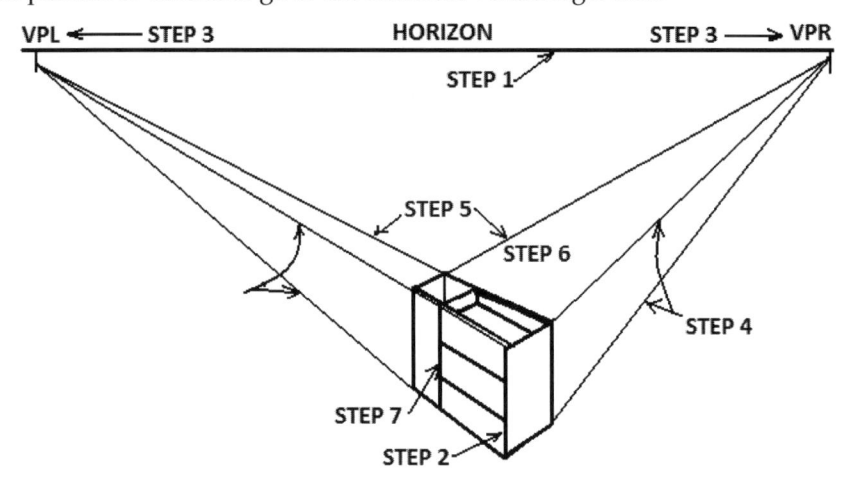

Figure 8-12: Angular perspective drawing

Construction of a perspective drawing

Step 1: Locate a suitable station point (SP). The distance between the station point and the object represents the true distance from the viewer to the building to the scale of the drawing. Accordingly, the longer the distance the smaller the building will appear in the picture.

Next draw a center line of vision i.e., a line from the station point to the building. Fix the drawing on the drawing board with the center line of vision (CLV) in a vertical position and cover with a transparent paper. Check that the building is falling within a 60° cone of vision, since parts of it falling outside this cone will appear distorted when looking at the picture.

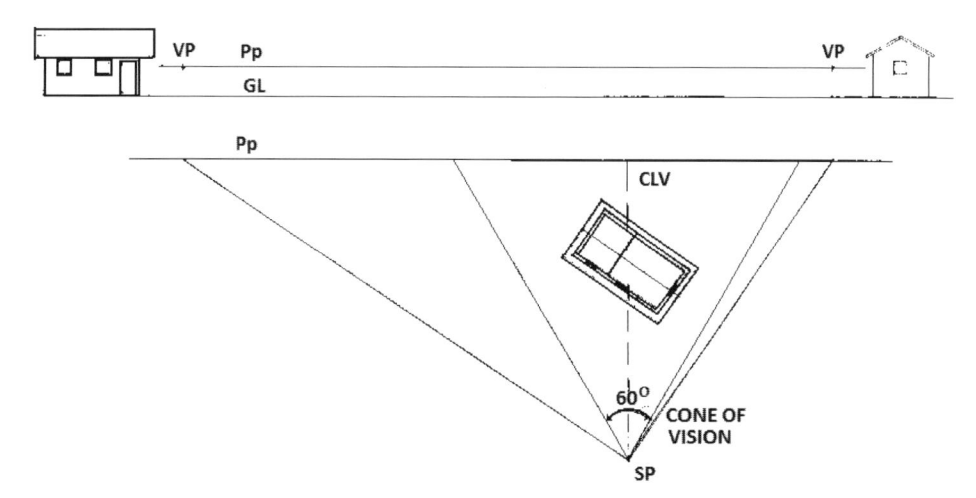

Figure 8-13 Construction of a perspective drawing

Step 2: Locate the picture plane (PP) and vanishing points (VP). The picture plane is a line drawn at 90° to the center of vision line i.e., horizontal on the drawing board. The distance between the station point and the picture plane will directly influence the size of the perspective picture. Think again of the situation where the outline of a building was traced on a window pane. If the window pane was moved closer, the outline picture would be smaller.

Thus, if the reader of a perspective drawing is to get an image of the true size of the illustrated building, he will have to look at the perspective from the same distance as the distance between the station point and the picture plane when it was constructed. Therefore this distance is normally taken to be 400 to 600mm. The vanishing points are then located by drawing lines from the station point to the picture plane parallel to the visual sides of the building.

Step 3: Locate the horizon line (HL) and the ground line (GL). The horizon line can be located anywhere on the paper as long as it is parallel to the picture plane, but leaving enough empty space to allow the perspective picture to be constructed around it. The ground line is then drawn parallel to the horizon line at a distance corresponding to the eye level to the scale of the drawing. The horizon line will always be above the ground line if the view point is above ground level. The vanishing points are then vertically transferred to the horizon line. It is helpful to put needles in the vanishing points on the horizon line to guide the ruler in further construction of the perspective.

Step 4: Locate a height line (HL) and mark the heights on this line. True heights of the building can only be scaled on a height line in the perspective picture. Start by locating a height line on which heights concerning the front wall can be scaled. This is a vertical line from the point on the picture plane where it is crossed by a line extended from the front wall in the plan view. The point where the height line crosses the ground line will represent ground level and all heights in the front wall can now be scaled from this point to the scale of the plan view. Top and bottom lines for the front wall can now be drawn from the vanishing point through the marks on the height line.

Step 5: Visual rays (VR) to locate points in the perspective view. Visual rays are drawn to locate the exact position of the corners of the front wall in the perspective. The rays are drawn from the station point through the point to be located in the perspective to the picture plane. From the picture plane the line is continued vertically to the intersection with the top and bottom lines. With further visual rays the outline of the visual walls can be drawn in the perspective.

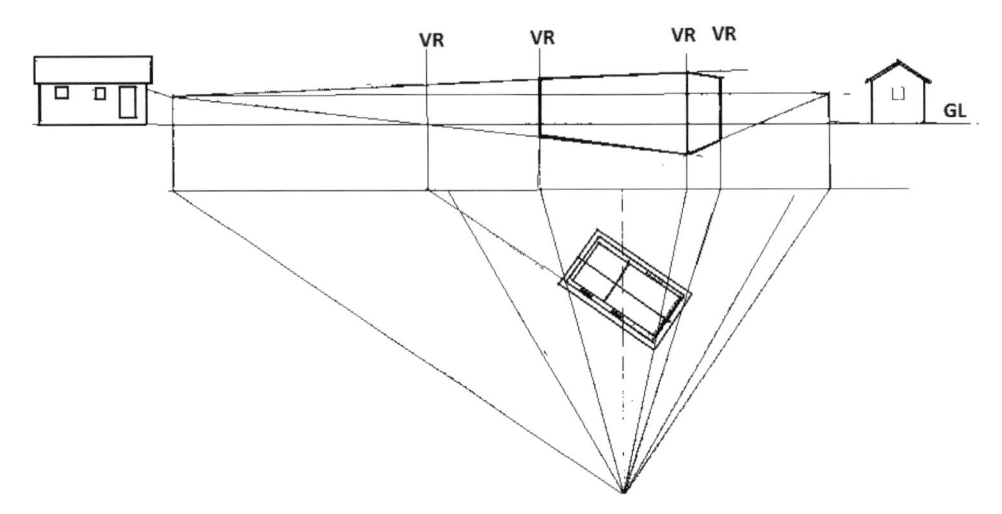

Figure 8-14: Visual rays (VR) in the perspective view

Step 6: Further height lines and visual rays. To find the top line for a double pitched roof a new height line must be constructed since that height is at a plane behind that of the front wall. Visual rays are then used to find the ends of the ridge. Doors and windows in the front wall are constructed with the height line for the front wall and further visual rays to find points in the perspective.

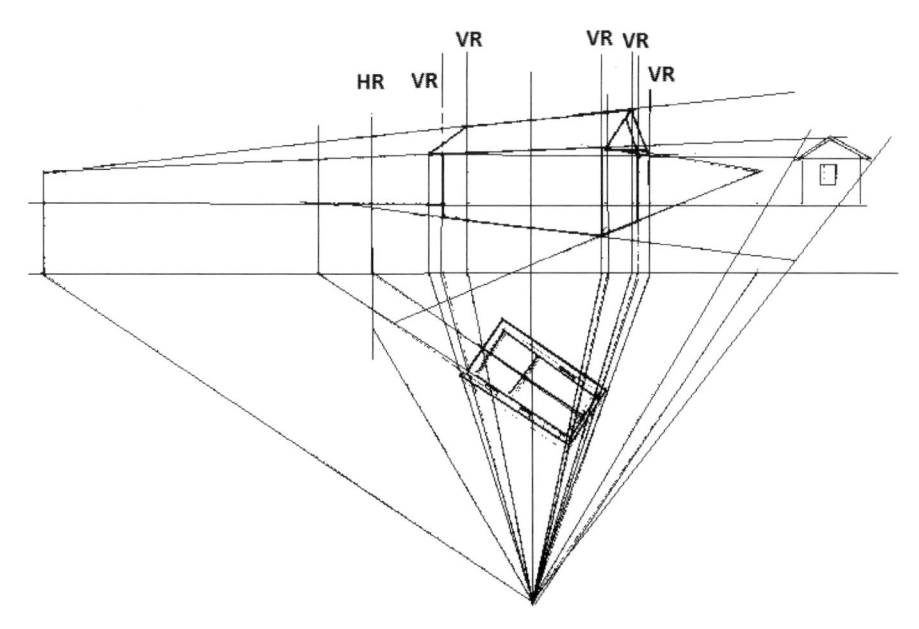

Figure 8-15: Locating Visual rays (VR) points

Step 7: Completing the perspective view. When the major outline of the building and principal objects in the visual sides, such as doors and windows, have been constructed in the perspective view, the drawing tends to be quite crowded with lines. Further details are therefore usually more easily constructed freehand.

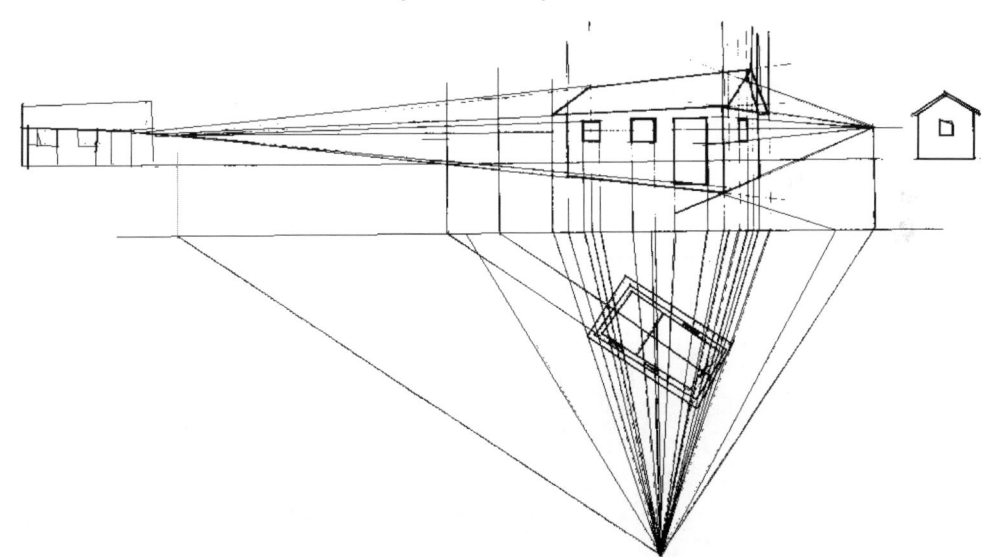

Figure 8-16: Tracing out the perspective

Finish the perspective by drawing vegetation and miscellaneous objects which will appear in the surroundings of the building. People in the picture will always be drawn with their eyes on the horizon line. The size will then determine the distance to the viewer. Finally cover the perspective drawing with tracing paper and redraw the picture leaving out all the construction lines.

Figure 8-17: Final perspective drawing

8.7 Diametric drawing

In diametric drawings, all dimensions along two axes are drawn to *true size*. The dimensions along the third axis are *halved*. This projection is preferred when one view of the object is to be emphasized than the other two views (i.e. when that one view is of more interest than the other views). Example includes a cube of length L.

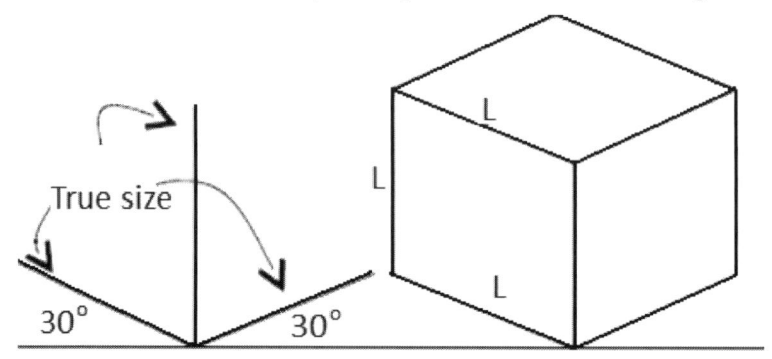

Figure 8-18: True sizes of an diametric box

8.8 Model building drawings

You need considerable experience to be able to visualize fully a building from a set of drawings. The farmer not only finds it very difficult to understand simple plan view and section drawings, but found it hard to interpret fully rendered perspectives. However, three-dimensional model drawings can be viewed from all sides bringing more realism to the presentation and this usually results in communication and transfer of ideas.

There are three types of models in common use for presentation of farm building projects:

1. Three-dimensional maps or site plans are used to present development plans for large areas or the addition of a new building on an old site with already existing structures. These models have contours to show the topography while structures are carried out in simple block form with cardboard or solid wood, usually without any attempt to show detail.

2. Basic study models are used for examination of relationships and forms of rooms and spaces in proposed buildings. They are often built in cardboard, and there is usually little attempt to show details, although furnishings and equipment may be indicated. Windows and door openings are shown with dark coloured areas or left open. Contours are shown only if they are of importance for the building layout.

3. Fully developed models may be used in extension campaigns, for public exhibition, etc. These models show details to scale and have close representation of actual materials and colours. Part of the roof is left out or made removable in models aiming to show the interior of a building.

The size of the model is determined by the scale at which it is made and the size of the actual project. While detail is easier to include in a model made to a large scale, too much detail may distract from the main outlines and essential features; and if too large, the model will be more costly and difficult to transport.

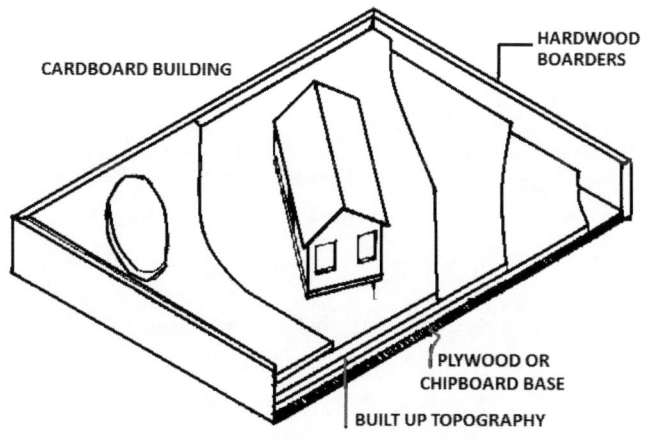

Figure 8-19: Typical study model

Basic study models are often made to a scale of 1:50 or 1:100 to allow for coordination with the drawings, while fully developed models of small structures may be made to a scale of 1:20 or even larger. Whatever scale is used for the model, it is desirable to include some familiar objects, such as people or cars, to the same scale as the model to give the observer an idea

Models can be increased in strength and rigidity by bracing the walls with square pieces of cardboard in positions where they will not show in the finished model. Bracing is particularly important in models which are going to be painted as paint will tend to warp cardboard and sheet wood if applied over large areas. Regardless of the material being represented, colours should be subdued and have a flat, not glossy, finish. Distemper or water colour is best for use on cardboard and unsealed wood, but care must be taken to remove excess glue as this will seal the surface and cause the colour to peel off.

A photograph of the model may be used in cases where it is not feasible to transport the model or when photos need to be included in information material and the actual building has not yet been completed. Models often appear more realistic when photographed, particularly in black and white, because of better contrast, but adequate lighting from a direction which produces a plausible pattern or sun and shadow on the building must be assured. Outdoor photography allows for a sky or terrain background to be incorporated into the photograph of the model.

8.9 Assembly drawings

The previous chapters covered the general aspects of engineering drawing and how to produce a detailed drawing of a single part with all the necessary information to make the part. The assembly of these parts is shown in an assembly drawing also known as a general arrangement.

Features of an assembly drawing

Dimensions: Detailed dimensions required for manufacture are excluded from assembly drawings. But overall dimensions of the assembled object are usually indicated.

The spatial relationship between parts that are important for the product to function correctly should also be indicated on the drawing for example indication of the maximum and minimum clearance between two parts.

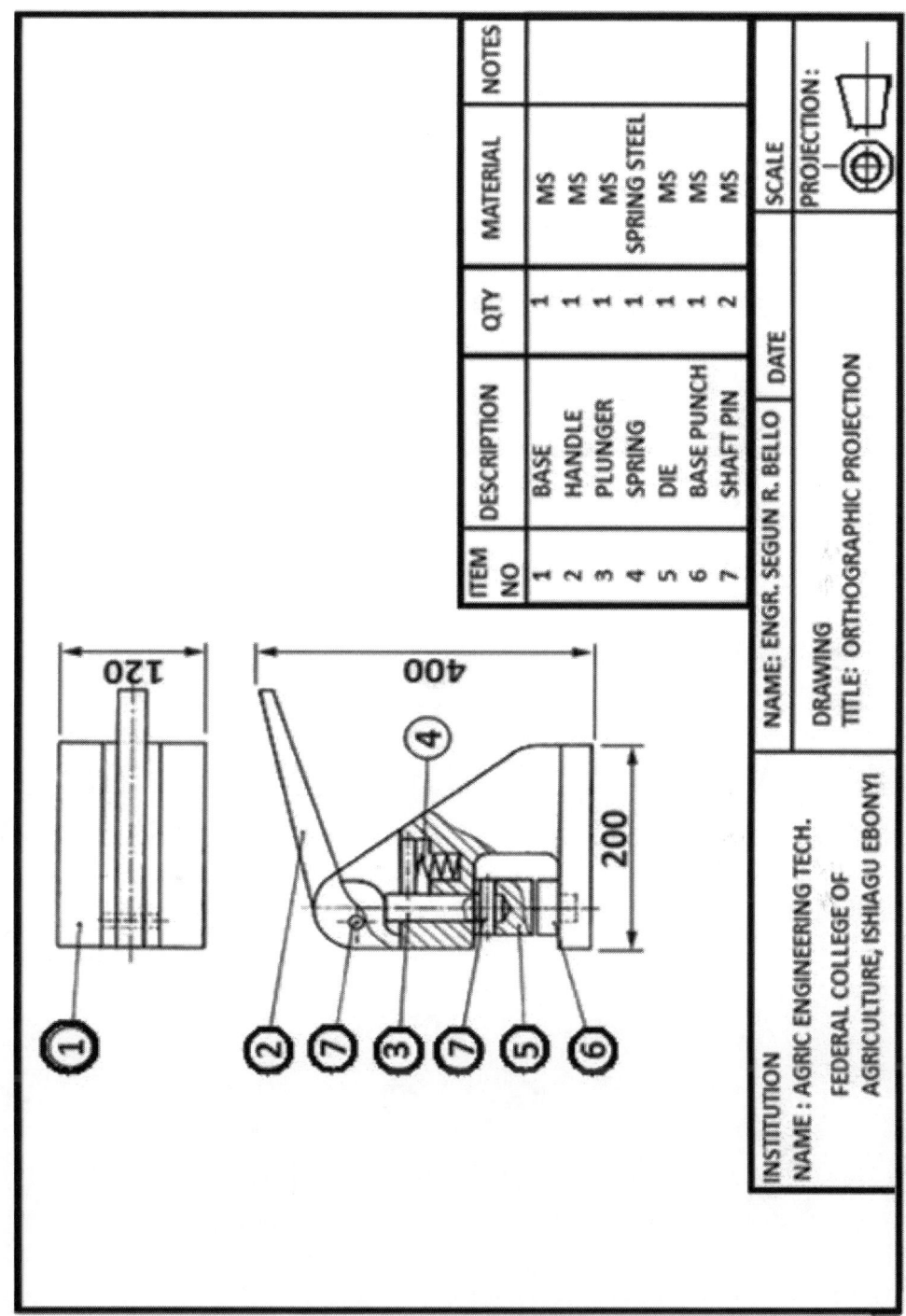

Internal parts: If there are internal assemblies, sectional views should be used.

Figure 8-20: Drawing an assembly drawing

Parts list: Each part is given a unique number, indicated on the drawing by a circle with the number in it and a leader line pointing to the part. The leader line terminates in an arrow if the line touches the edge of the component or in a circle if the line terminates inside the part.

A table of parts should be added to the drawing to identify each part, an example of a parts list is shown below:

Item No.	Description	Qty	Material	Remarks

The first three items; Item No., Description, and Quantity should be completed for every distinct part on your drawing (i.e. the numbers of duplicate parts are recorded in the quantity). The material is used for components that are being made within the company. The Remarks column is useful for specifying a manufacturer's part number when using bought-in parts.

8.10 Sketch drawings

Free hand sketching is a very useful and widely used form of 'technical' communication. The idea is to accurately communicate the required information. This does not necessarily need artistic ability! Simple freehand sketches are convenient forerunners to final working frequently used for preliminary studies or to illustrate an explanation during a discussion. They are also the logical way for the building designer to convey his ideas to the draftsman.

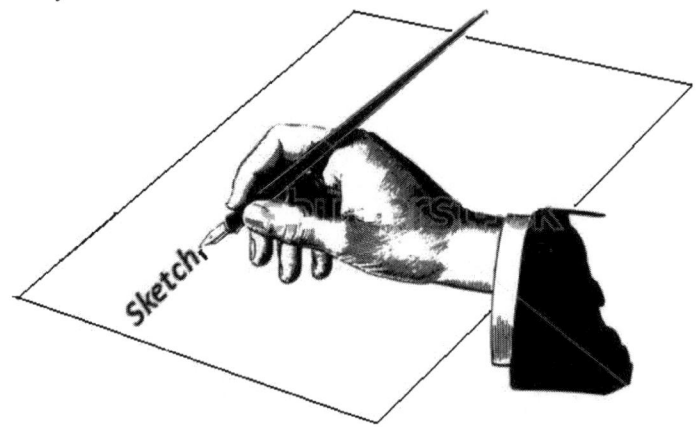

Figure 8-21: Free hand sketch

Free hand sketching may be used for developing plans by testing a number of alternative designs or for evolving detail drawings of complex building elements. They are particularly useful in recording details and dimensions from existing structures or prefabricated units. The ability to make use of free hand without the aid of drawing instruments, as a guide, in sketching is a valuable asset acquired by constant practice.

Importance of sketching

As an engineer you need to be able to produce freehand sketches for various reasons, such as:

1. To help visualize your own ideas
2. To convey your ideas, information and details to others
3. To record details of measurements for later use
4. To help produce a working drawing
5. Freehand sketching makes student observant and accurate.

Sketching requirements

Any sketch which is made should satisfy the following requirements.

a. It should describe the shape of the object completely, showing the relative parts in fair proportion but not to any particular scale.
b. It should carry all essential dimensions.
c. It should have notes to specify, for example, the material of which the object is made and the method of manufacture.

Rules for making freehand sketching

Principal lines are sketched lightly using a number of short strokes. Once the joining points have been established and lines are satisfactorily straight, they may be darkened as needed to give emphasis and easy reading. Although they are not given a scale but only needed to be in approximate proportion, all measurements should be clearly shown with dimension lines and legible figures and symbols.

Just as with final drawings, plan (top) and section (front + side) views are simplest to sketch and dimension. However, isometric sketches are useful in presenting a more pictorial view of a structure.

The rules or procedure for making sketches are few.

1. Straight lines are best drawn lightly in pieces and then made continuous. See Figure 1

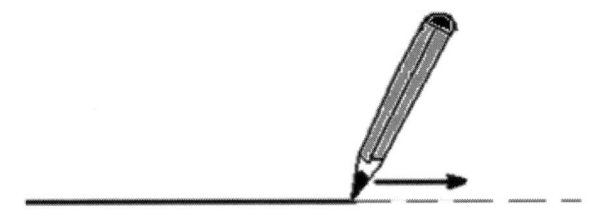

Figure 8-22: Free-hand drawing of straight line

2. Sketching circles require a few marked diameters. The bare outlines of the views should be spaced out in thin lines, and center lines should be inserted. The circle or part of a circle is easier to draw if split into quarters. Sketch lightly before lining in.
3. Other details can be added and then followed by the final thickening in and the addition of dimensions.

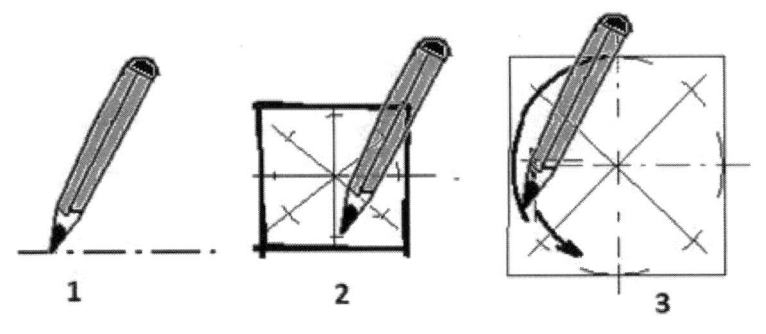

Figure 8-23: Free-hand drawing of curved line

4. For pictorial views, the three isometric axes, (x, y, z) should be drawn, together with, if necessary, the rectangular frame which would hold the object. NB. This will be made clearer in the topic.
5. If it occurs to student that a sections view would make the figure or object clearer, it can be added.

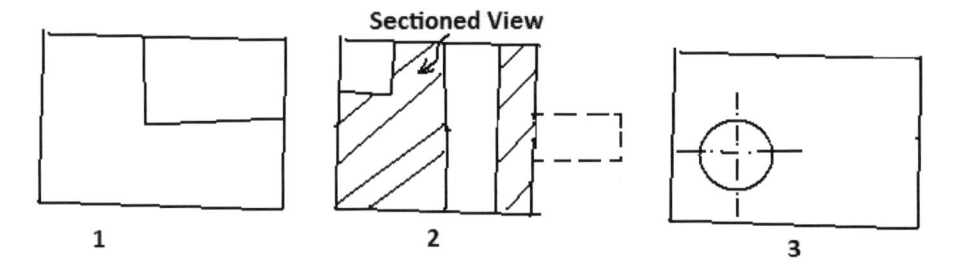

Sectioned View

Figure 8-24: Free-hand drawings

Note to draw the protecting cylinder we must concentrate on the hidden face and locate the position of the circle on that face as shown in figure below

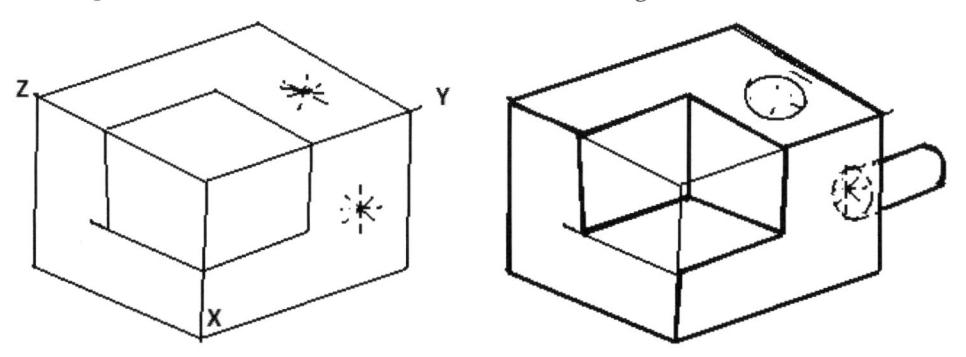

Figure 8-25: Free-hand drawings

6. Always use an H or HB pencil for sketching. Note that many objects are symmetrical i.e. they are the same shape each side of a center line. Take for example, sketching a chisel tool. The stages are as stated below:

   a. Observe the length in relation to the width, mark out a box with a center line.
   b. Mark off the other proportions. Lightly sketch in the curves
   c. Line in with short accurate strokes.

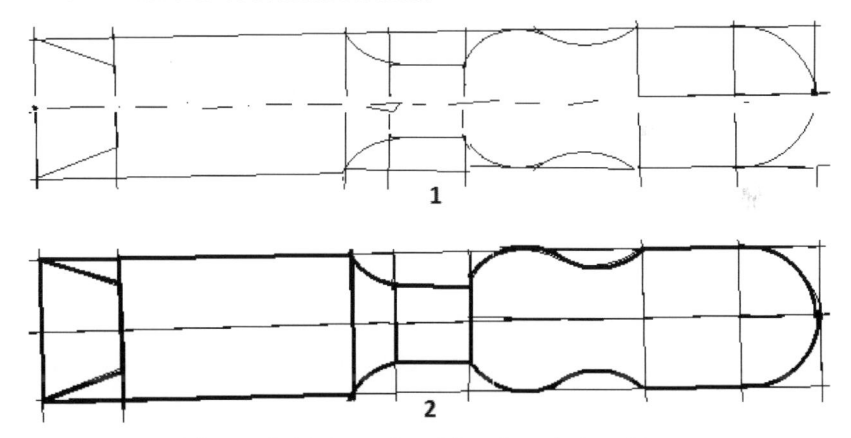

Figure 8-26: Sample of free-hand sketching

In the same way, polygons, quadrilaterals etc. can be sketched, for instance, a regular polygon on flat base (Figure 8-27). You must first have to start with a good estimation of the angle 60°.

Next sketch the figure following these instructions:

1. draw a square that will contain the figure and mark out the sides figure 1
2. Join the sides and complete the octagon figure 2.

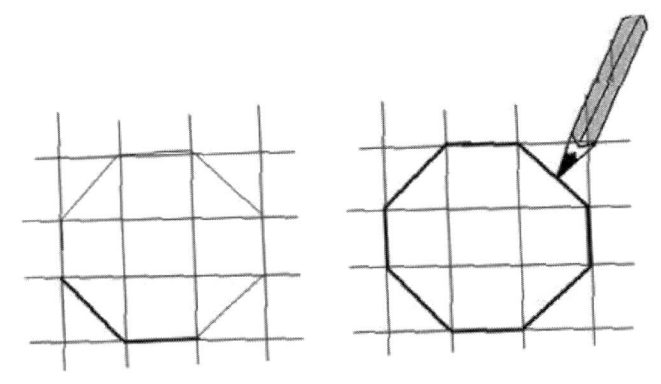

Figure 8-27: Sketching polygon in grids

General tips for sketching straight lines

Short horizontal lines, say about 150mm or less, can usually be drawn using a single stroke. It may help to put a short dash at the two ends of the intended line and then join them together. Initially it may also help to practice drawing the line a few times, without actually contacting the pencil with the paper. Longer lines can be drawn as a series of short-stroke lines overlapping. Direction from left to right is probably easiest if you are right handed.

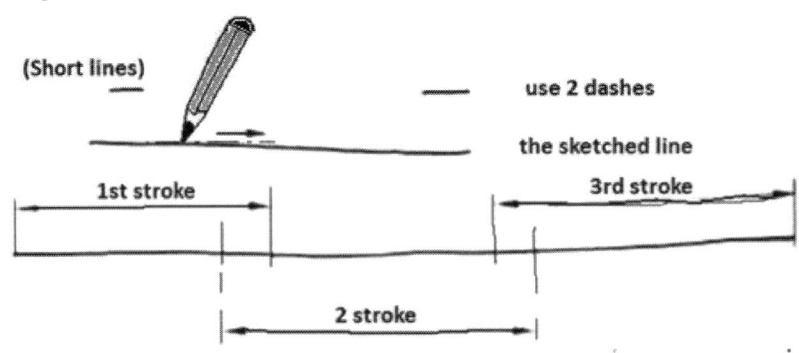

Figure 8-28: Stroking straight lines

Vertical lines tend to be drawn more easily using movement of the fingers rather than the hand. Again longer lines can be made up by overlapping shorter lines. With practice, longer continual vertical lines can be drawn by moving the whole hand.

General tips for sketching curves and circles

The secret to sketching circles, or curves, is in the creation of construction lines to guide the line of the curve at critical points. For example, consider sketching a circle. Using the knowledge of circles symmetry construct centerlines and a bounding box. Start drawing the circle by drawing the arcs at the centerlines, then complete. Another method could be to sketch more arcs around the center to build up the circle shape more gradually.

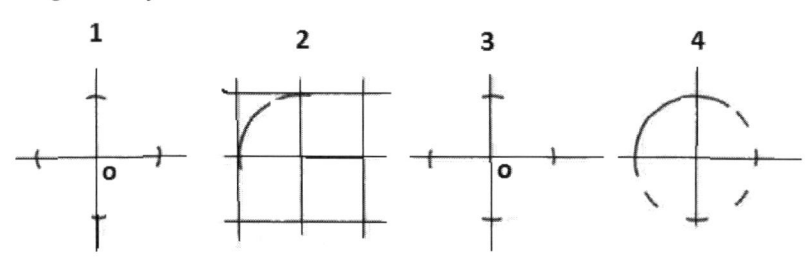

Figure 8-29: Sketching curves and circles

Sketching orthographically

When a component or building structure is to be sketched, it is a good idea to divide it up into simplified shapes, such as squares and rectangles, to use as guidelines. This will also help to maintain the correct proportion of the sketch. Use one of the smaller features of the object being sketched as a reference for the relative sizes of the other features.

Scale itself is not important in sketching. For example, creating a simple orthographic sketch of the wall shown on the next page, you could split the features up as shown and use the proportions to create the construction lines for the orthographic sketch, which has been started.

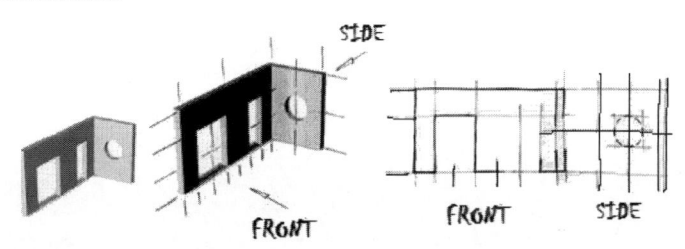

Figure 8-30: Orthogonal sketching

Sketching pictorially

A similar approach can be used when creating perspective, isometric or oblique projection views as well. Construct basic shape guidelines and then add the detail, using a smaller feature as a guide for proportion. A sketched isometric view of a block is shown in isometric gridlines in Figure 8-30 while Figure 8-31 show an orthographic view drawn on grid points.

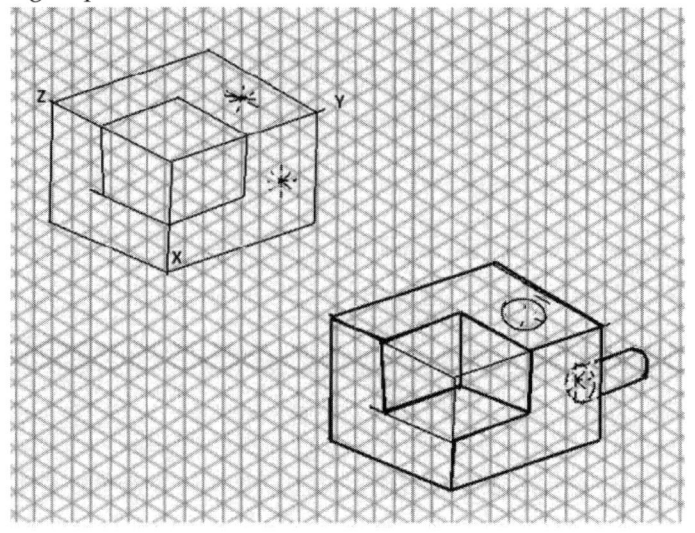

Figure 8-31: Sketching on isometric grid

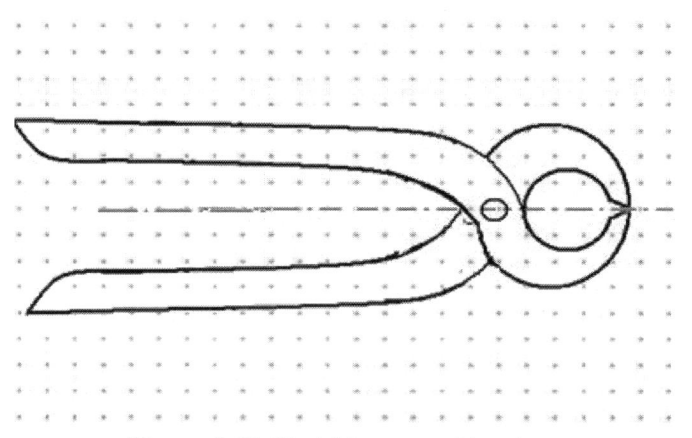

Figure 8-32: Sketching on grid points

Sketching aids

A soft pencil, eraser, inexpensive paper and a clipboard complete the sketcher's equipment. When a final design has been chosen, it is drawn with instruments on tracing paper so that prints may be readily made. A 70/75g paper is usually sufficient. However, if many prints are to be made a heavier paper should be used.

Printing of sketches

Plastic tracing film is a new material which is more durable for handling and storage and has the advantage that ink can be removed with a moist eraser. It is however much more expensive than tracing paper and requires the use of special lead and drawing pens, since its surface is much harder.

Whatever paper is chosen, it is best to use drafting tape to affix it to the table as the low adhesion allows easy removal without damage.

8.11 Comparing projections

Isometric and orthographic projections

The following differences exist between isometric and orthographic drawings

S/No	Orthographic	Isometric
1.	Line drawing	Pictorial drawing
2.	Orthogonal lines	Inclined lines
3.	Drawn at 90 degrees to horizontal plane	Drawn at 30 degrees to the horizontal plane
4.	Suitable for engineering details	Suitable for pictorial presentation
5.	Carries dimension lines	Dimension lines not required
6.	Accommodates hidden details	Hidden details not required except for purposes of teaching

Isometric and oblique projections are useful in presenting a pictorial view of a structure and are particularly suitable for free-hand sketching, although the views may be slightly distorted.

Axonometric projection is best suited to show the interior of rooms with its furniture, equipment or machinery.

The *perspective projection* is a bit more complicated to construct, but gives a true pictorial view of a building as it will appear if standing at about the same level as the building and at some distance.

The *most realistic* of all the pictorial drawings is the perspective drawing because of the followings;

1. Receding lines in the drawing "meet" at a vanishing point instead of being parallel
2. Eliminates distortion at the back part of pictorial drawings

Advantages of oblique drawing over orthographic

1. An oblique drawing is useful when the front contains more details and features than the side view
2. A mental image can be created more quickly than with orthographic alone

Exercise

1. Draw the layout drawing (only) of the instruments below

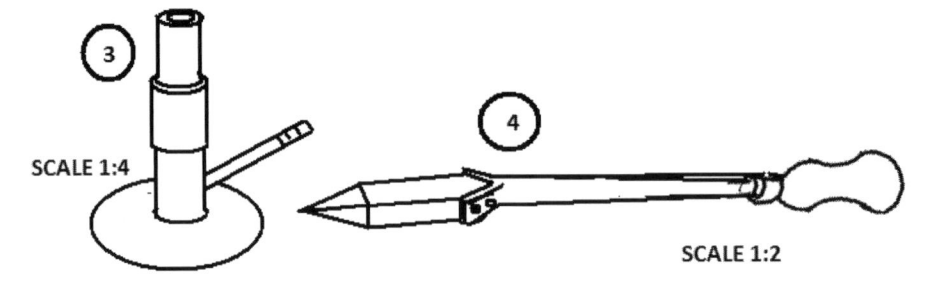

   SCALE 1:4

   SCALE 1:2

2. The drawing of a door catch is shown below, draw a free hand sketch to given proportion and dimensioned appropriately.

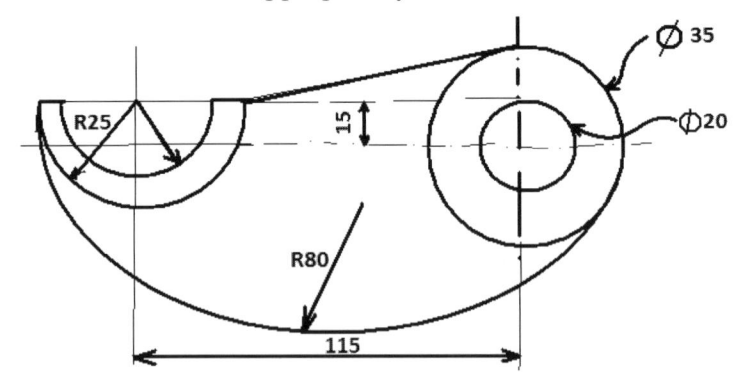

3. Using free hand. Show clearly the method used to produce the curved parts.

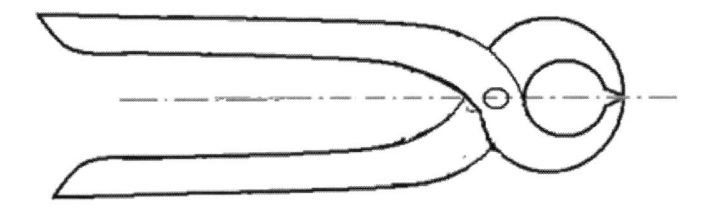

4. Use the Isometric grid lines on the right hand in the figures below to sketch the clamp and block shown below. Add shading lines to make it look more realistic

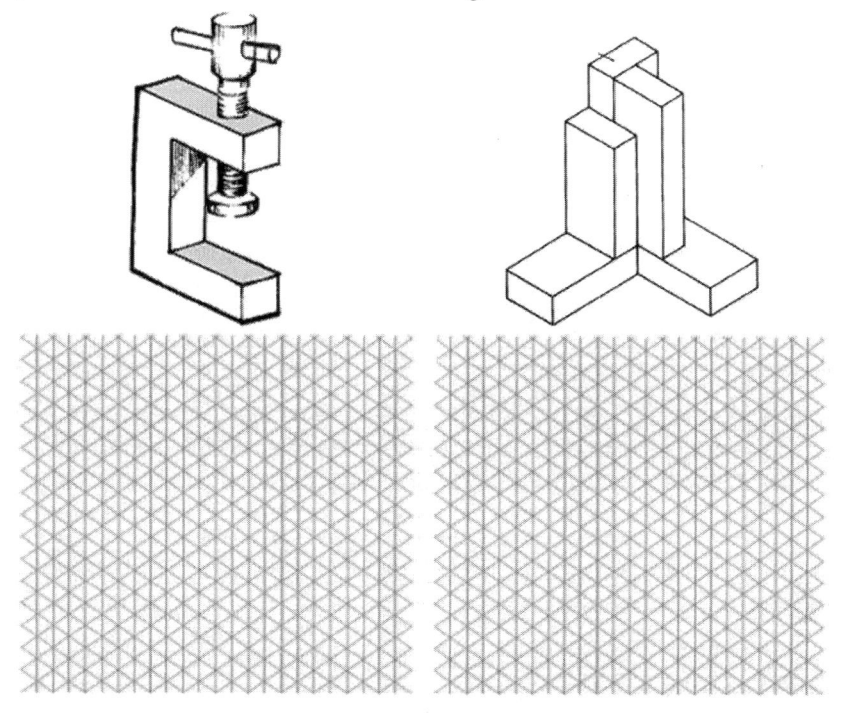

5. The most important rule in freehand sketching is _____ the sketch in _____. No matter how brilliant the technique or how well the small details are drawn, if the _____ especially the large overall _____ are bad, the sketch will be bad.
6. Produce to scale the isometric view shown in figure below

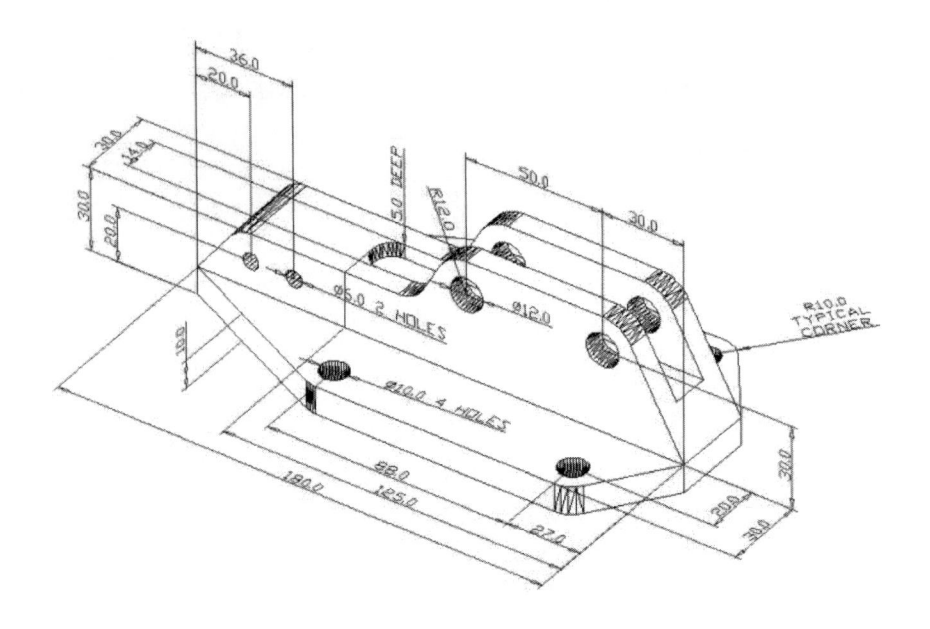

SECTION 3
Conic Sections & Surface Development

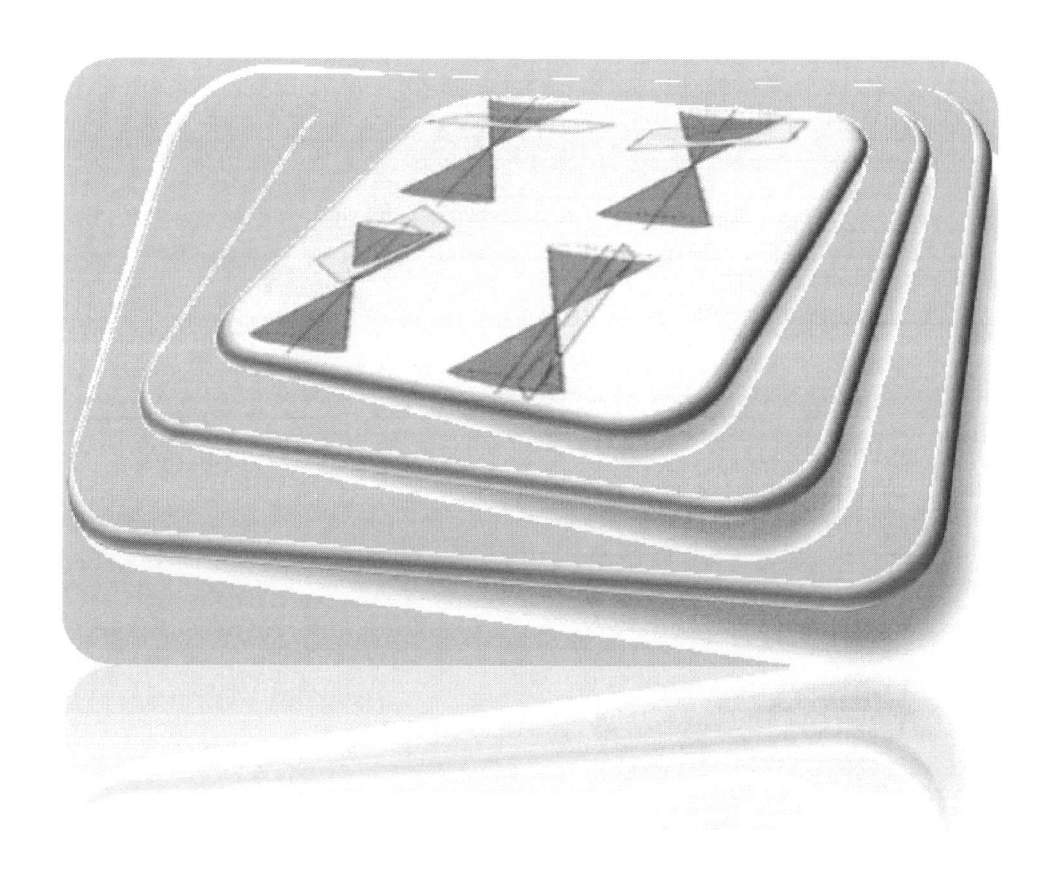

Introduction

The basic shapes met with in practice are usually flat cylindrical or conical. These basic shapes include ellipse, parabola and hyperbola which are called conic sections because these curves appear on the section of a cone when it is cut by some typical cutting planes.

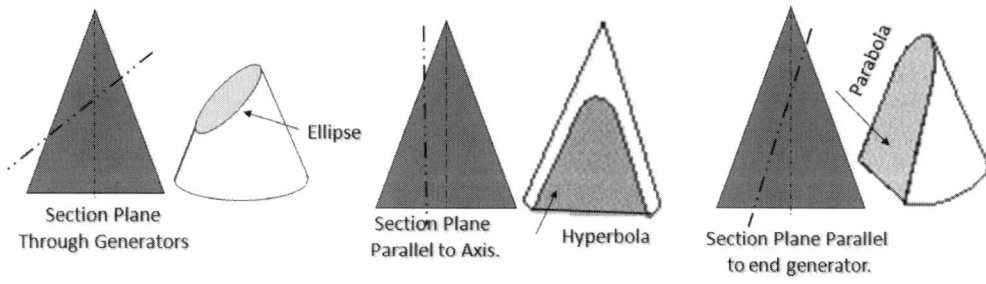

Examples of conic sections

This chapter concentrates on conic-specific descriptions, methods of constructing conic curves, given points and other details.

At the end of this chapter, students are expected to be familiar and be able to

a. Identify the general shape of a conic section (circle, hyperbola, ellipse, parabola) produced by a given equation
b. Draw sectional views of simple solids
c. Determine the true length of lines in space
d. Draw true shape of sections
e. Construct intersection of solid and solids such as prisms, pyramids, right circular cones.
f. Construct intersection of solid and plane
g. Make surface development of prisms, right cylinder, right cone, trays and simple transition pieces.
h. List practical examples of surface development
i. State and explain the applications of auxiliary projections
j. Project first auxiliary plain and elevation from given principal views.

Chapter 9

Conic Sections and Construction

9.1 Introduction

Conic sections are curves which can be derived from taking slices of two cones "nose to nose", with one cone balanced perfectly on top of the other (Figure 9-1). "Section" here is used in a sense similar to that in medicine or science, where an extremely thin slices ("sections") are shaved off from sample (a specimen, for instance) for viewing under a microscope.

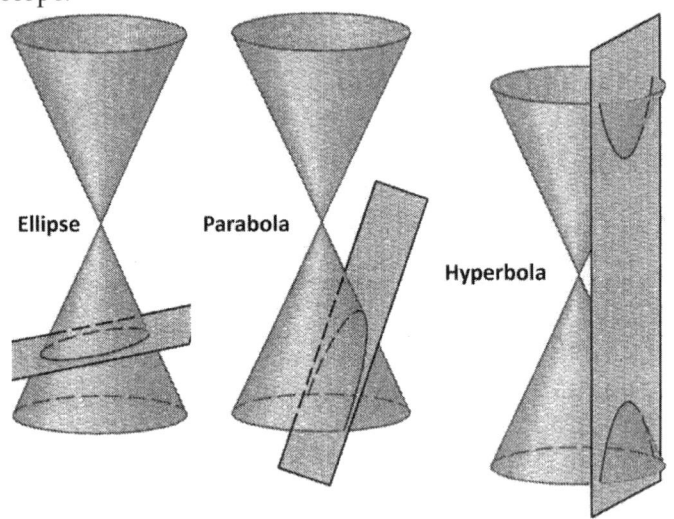

Figure 9-1: Conics

A conic section may be defined as the locus of a point P that moves in the plane of a fixed point, F, called the focus and a fixed line D called the conic section directrix such that the ratio of the distance P from F to the distance F to D is a constant, e, called the eccentricity. A conic section is a curve described by a point which moves in a plane in such a manner that its distance from a fixed point in the plane (a focus) is in a constant ratio to its distance from a fixed line (a directrix) in the plane.

Basic terms and definitions

There are some basic terms used in conic section drawings which are defined as follows:

- *Directrix*: This is a line AB (Figure 9-2) from which distances is measured in forming a conic; the plural form is "directrices".

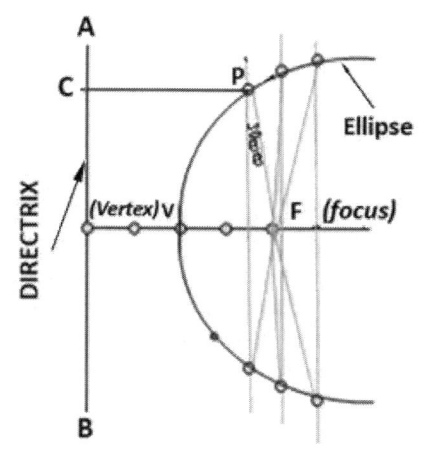

Figure 9-2: Features of conic sections

- *Eccentricity*: This is the ratio of the distance from a fixed point in a plane (a focus, F) to its distance from another fixed line (a directrix AB) in the plane (Figure 9-2). This ratio is known as the eccentricity (E). This ratio is expressed mathematically as:

$$E = \frac{PF}{PC} \quad \ldots \ldots \ldots .1$$

Figure 9-4 below shows the effect of change in eccentricity on the shape of different curve when a constant focus, F and directrix DD_1 are used.

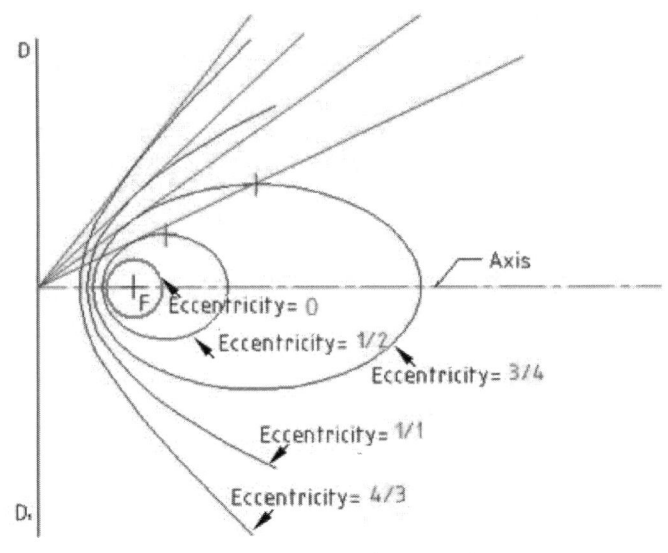

Figure 9-3: Lines of eccentricity

The curves in the figure are circle, an ellipse, a parabola, or a hyperbola, according as their eccentricity equals zero, less than, equal to, or greater than unity (one) respectively as illustrated below.

1. For a circle   E=0
2. For Ellipse   E is less than unity (E=3/4: E<1)
3. For Parabola   E is equal to unity (E=1/1: E=1)
4. For Hyperbola   E is greater than unity (E=4/3: E>1)

By implication;

   a. Only when the eccentricity is unity is the curve a parabola, when the eccentricity rises over unity the curve becomes a hyperbola
   b. As the eccentricity approaches zero the ellipse becomes more circular
   c. The angle between the line of eccentricity and the axis of symmetry will always be less than 45° for an ellipse, exactly 45° for a parabola and greater than 45° for a hyperbola and
   d. The higher the eccentricity the straighter the curve becomes.

- *Focus*: This is a point along the major axis from which distances along the curve are measured in forming a conic. It is a point at which these distance-lines converge. The plural form of focus is "foci".
- *Vertex*: Vertex is the point of inflection or turning point along a curve. In the case of a parabola, the point (h, k) at the "end" of a parabola is called vertex; in the case of

an ellipse, it is located at the end of the major axis (Figure 9-4); in the case of hyperbola, it is the turning point of a branch of the hyperbola. The plural form of vertex is "vertices".

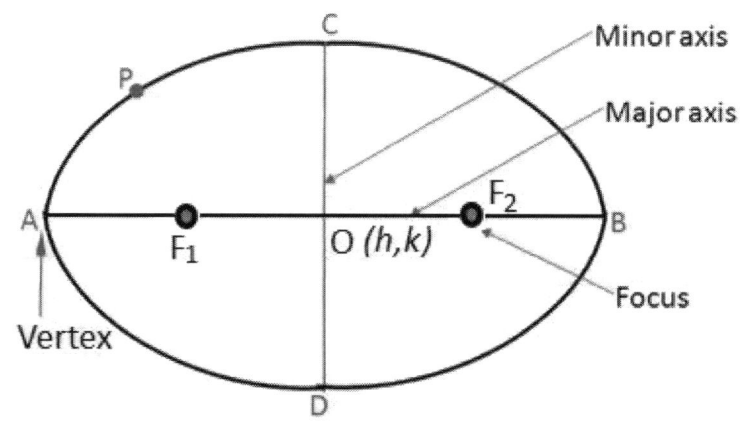

Figure 9-4: Features of conic (ellipse) sections

- *Center*: This term refers to a point O (h, k) at the center of a circle, an ellipse, hyperbola or a parabola.
- *Axis*: This is a line drawn perpendicular to the directrix passing through the vertex of a conic. It is also called the "axis of symmetry". The plural form is "axes". There are two categories of axis; major and minor axes.
- *Major axis:* A line segment perpendicular to the directrix of an ellipse and passing through the foci; the line segment terminates on the ellipse at either end; also called the "principal axis of symmetry"; the half of the major axis between the center and the vertex is the semi-major axis.
- *Minor axis:* A line segment perpendicular to and bisecting the major axis of an ellipse; the segment terminates on the ellipse at either end; the half of the minor axis between the center and the ellipse is the semi-minor axis.
- *Locus:* A set of points satisfying some condition or set of conditions; each of the conics is a locus of points that obeys some sort of rule or rules; the plural form is "loci".

9.2 Construction of conic sections

The methods of construction of the three curves; ellipse, parabola and hyperbola are identical. Some of these methods are considered bellow.

9.2.1 Ellipse and its construction

Ellipse defined

When a point moves so that the sum of its distances from two fixed points, called focal points or loci, is constant, then the locus of the points is an *ellipse*. The constant is the major axis of the ellipse. Ellipse shapes are usually found in arches, bridges, dams, monuments, man-holes, glands

Basic geometrical properties employed in constructing an ellipse

Some basic geometrical principles of guiding ellipse construction include:

1. *Equation*: The equation of an ellipse is expressed as

$$\frac{x^2}{a^2} + \frac{y^2}{b^2} = 1$$

Where

a: Half length of major axis, b: Half length of minor axis

2. *Eccentricity*: The eccentricity is expressed as:

$$E = \frac{PF}{PC}; \; E < 1$$

3. With reference to Figure 9-5 below, distances of points P, Q, R etc along the curve to focal points F_1, F_2, etc are designated as;

$F_1P = A1$, $F_2P = B1$

$F_1Q = A2$, $F_2Q = B2$

$F_1R = A3$, $F_2R = B3$

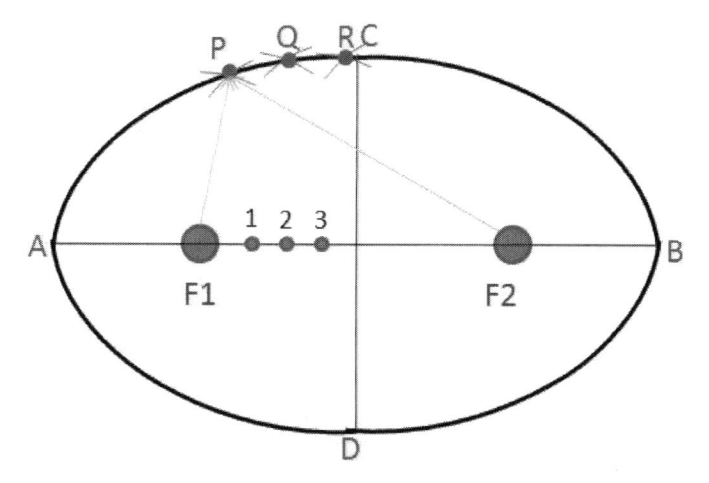

Figure 9-5: Features of an ellipse

4. The sum of distances of a point on the ellipse to the foci is constant; for instance

$$PF_1 + PF_2 = \text{Constant,}$$

$$QF_1 + QF_2 = \text{Constant}$$

5. The sum of distances from a point to foci equals the length of the major axis i.e.

$$AF_1 + AF_2 = AB = \text{Length of major axis}$$

6. The sum of the length of minor axis and the focal points F_1, F_2 is equal to twice the distance between one end of the minor axis and one of the focal point. By implication, this is equal to the height of the major axis. Expressed geometrically,

$$CF_1 + CF_2 = 2CF_1 = AB$$

$$\Rightarrow CF_1 = AB/2$$

Construction of an ellipse

Methods of construction of ellipse

Arcs of circles method

To construct an ellipse through arcs of circles method; given the length of major and minor axis you can always find the foci. You need the foci for some construction methods. Just draw radii of length from the ends of the minor axis.

Given the foci, however, you can't uniquely determine the axes. You need additional information such as the length of one axis. However, the major axis is always along the line through the foci and the minor axis always perpendicularly bisects the line between the foci.

Pin and string method

This makes use of the fact that an ellipse is the locus of points whose distances from two fixed points have a constant sum. The method is mostly useful as a demonstration tool to help students visualize ellipses, but it can be useful for constructing large ellipses.

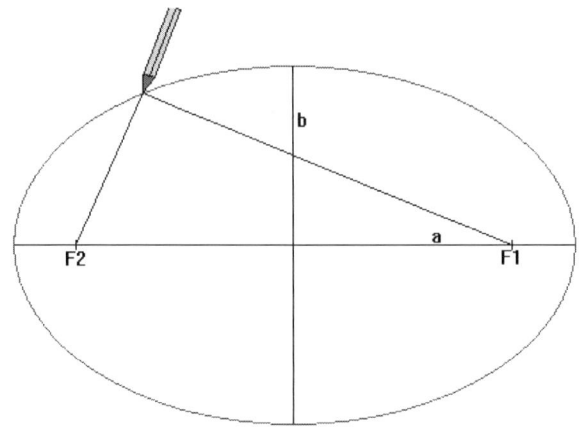

Figure 9-6: Pin and string method

Put a pin in each focus and tie a string to each pin leaving slack with length 2a. Pull the string taut with a pencil point and slide the pencil to draw the ellipse.

Directrix-focus method

To draw an ellipse given that the focus is 50 mm from the directrix and the eccentricity is 2/3.

Proceed as follows:

Note that from Figure 9-7, VE = VF1 and F1-P1=F1-P1' = 1-1'

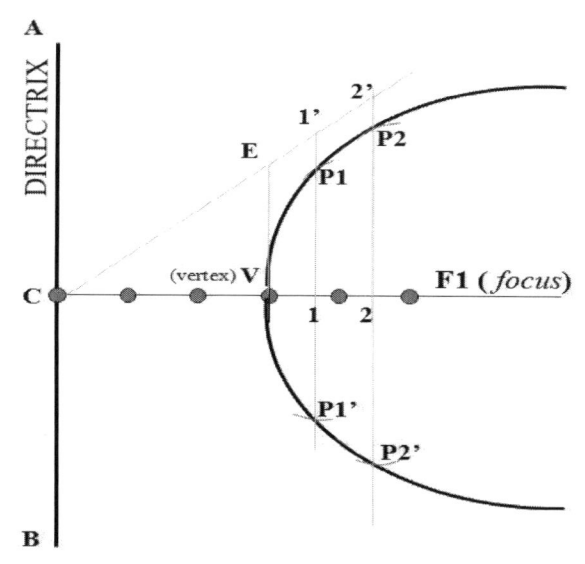

Figure 9-7: Ellipse by arc of circle method

F1-P1/ (P1 to directrix AB) = 1-1'/C-1=VE/VC (similar triangles)

=VF1/VC=2/3

Therefore P1 and P1' lie on the ellipse

Also, F1-P2=F1-P2'= 2-2' this implies that P2 and P2' also lie on the ellipse

Constructing an ellipse using the line of eccentricity

Assume the eccentricity of the ellipse intended is ¾; therefore the angle between the axis and the line of eccentricity is less than 45°.

Principle: For every point on the curve, the ratio of the distance to the focal point and the distance to the directrix is in a ratio of ¾ expressed as the eccentricity.

$$E = \frac{Distance\ to\ the\ focal\ point}{Distance\ to\ the\ directrix} = \frac{PF}{PD} = \frac{3}{4}$$

Hence proceed as follows to construct the ellipse:

Procedure

1. Given that the eccentricity is 3/4, measure out 4 units along the major axis and 3 units up perpendicular (minor axis) from the major axis. The line of eccentricity will come from the point where the axis meets the directrix and will pass through this point.
2. To get the vertices, project lines at 45° to the axis to meet the line of eccentricity, and then drop a perpendicular to the axis to locate the points.

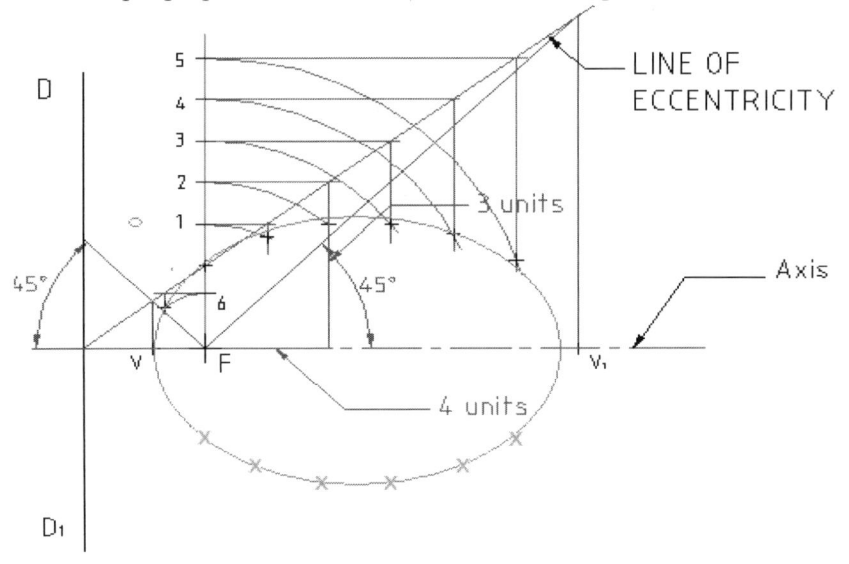

Figure 9-8: Eccentricity method of construction of ellipse

3. Then draw a line up from the focal point, where this line meets the line of eccentricity will be a point on the curve.
4. Divide the line into a number of parts, (for point 1) Take a line from the point parallel to the axis to meet the line of eccentricity, from this point drop a line parallel to the directrix.
5. With centre point F and your compass set to 1 swing an arc to meet the line parallel to the axis, this will be a point on the curve
6. Repeat steps 5 and 6 for the other points on the line.
7. When you have a sufficient amount of points on the curve you can sketch it in. (axial symmetry can be used to get the curve below the axis)

Construction of ellipse by concentric circle (envelop) method

In constructing an ellipse by concentric circle method, a series of lines are generated by the two given axes of the form represented in Figure 9-9 below.

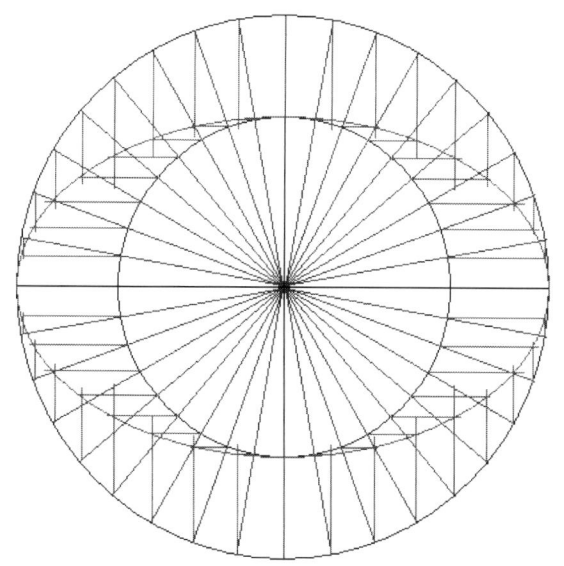

Figure 9-9: Constructing ellipse by concentric circle

For instance if it is required that an ellipse be drawn by concentric circles method given the major axis to be 100 mm and minor axis 70 mm long.

Proceed as follows

1. Draw both axes AB and CD as perpendicular bisectors of each other and name their ends as shown in Figure 9-10.
2. Taking their intersecting point as a center (0, 0), draw two concentric circles considering both as the respective diameters.
3. Divide both circles into 12 equal parts and label each point 1 to 12 as shown.
4. From all points of the outer circle draw vertical lines downwards and upwards respectively.
5. From all points of inner circle draw horizontal lines to intersect those vertical lines.
6. Mark all intersecting points properly as those are the points on ellipse.
7. Join all these points along with the ends of both axes in smooth possible curve as the required ellipse.

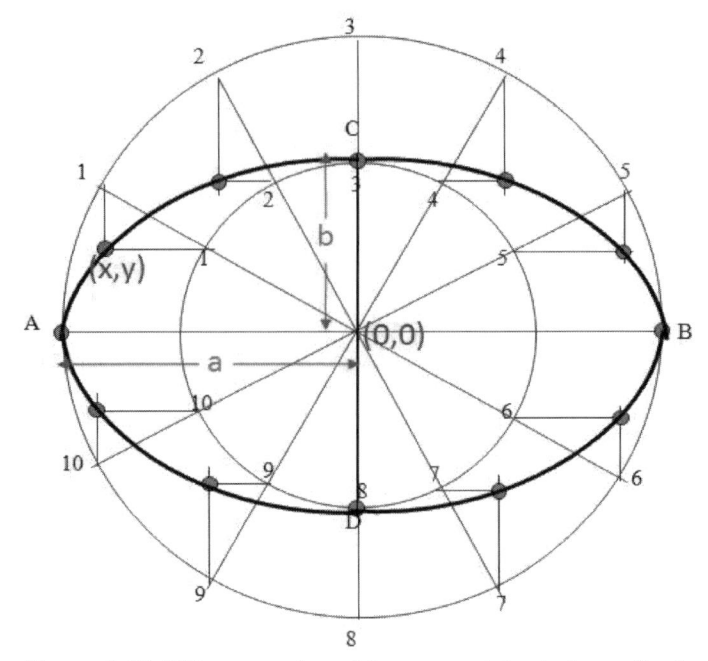

Figure 9-10: Ellipse produced by concentric circle method

Ellipse construction by Trammel method

To construct an ellipse by trammel method, proceed as follows

1. Draw the major and the minor axes perpendicular to each other.
2. With the major and minor axes constructed (and extended), mark a piece of paper with points O, A and B such that OA = a and AB = b.
3. Slide O along the minor axis and B along the major axis. Point A traces out the ellipse.

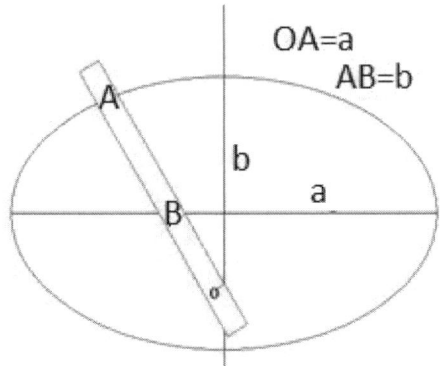

Figure 9-11: Trammel method (internal)

With any adequate care, this method is quite accurate and very fast. For cases where the axes are similar in size, the method above may be inaccurate.

It is also possible to draw an ellipse using an external trammel following the same principle (Figure 9-12).

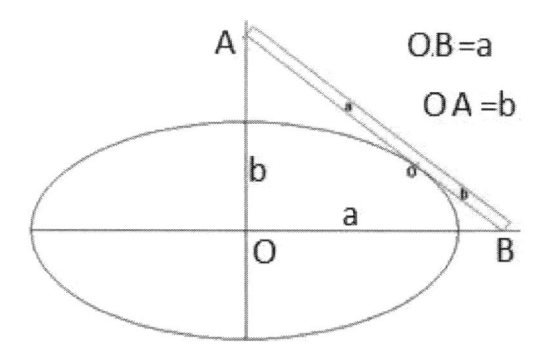

Figure 9-12: Trammel method (external)

Construction of ellipse by rectangle method

If it is required to draw an ellipse by rectangle method, taking major axis to be 100 mm and minor axis to be 70 mm long.

Proceed as follows

1. Draw a rectangle taking major and minor axes as sides.
2. In this rectangle draw both axes as perpendicular bisectors of each other..
3. For construction, select upper left part of rectangle. Divide vertical small side and horizontal long side into same number of equal parts.(here divided in four parts)
4. Name those as shown..
5. Now join all vertical points 1,2,3,4, to the upper end of minor axis. And all horizontal points i.e.1, 2, 3, 4 to the lower end of minor axis.
6. Then extend C-1 line up to D-1 and mark that point. Similarly extend C-2, C-3, C-4 lines up to D-2, D-3, & D-4 lines.
7. Mark all these points properly and join all along with ends A and D in smooth possible curve.
8. Do similar construction in right side part along with lower half of the rectangle.
9. Join all points in smooth curve. This is the required ellipse.

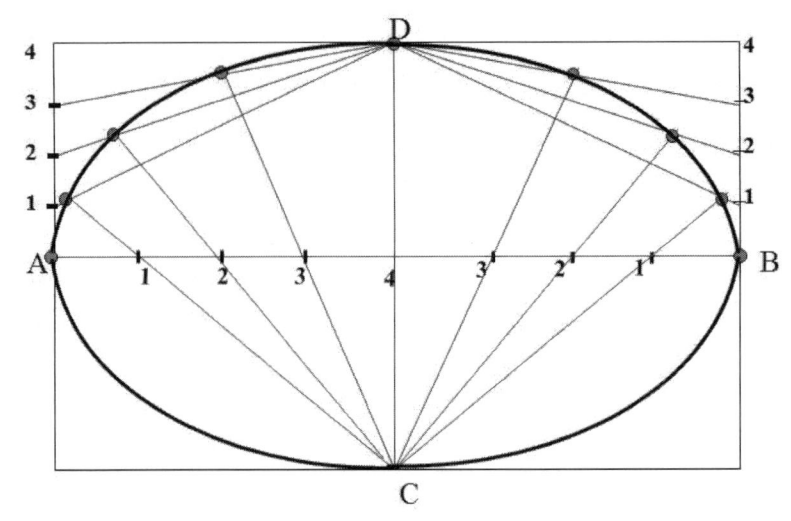

Figure 9-13: Constructing ellipse by rectangle method

Constructing ellipse by oblong/parallelogram method

If it is required to draw an ellipse by oblong method, of 100 mm and 70 mm long sides with included angle of 75° inscribed, the steps are similar to the previous case of rectangle method. Follow the processes below

1. Follow the procedures outlined for rectangle method only in replace rectangle, with the parallelogram. Draw the parallelogram of 100 mm and 70 mm long sides with included angle of 75° inscribed.

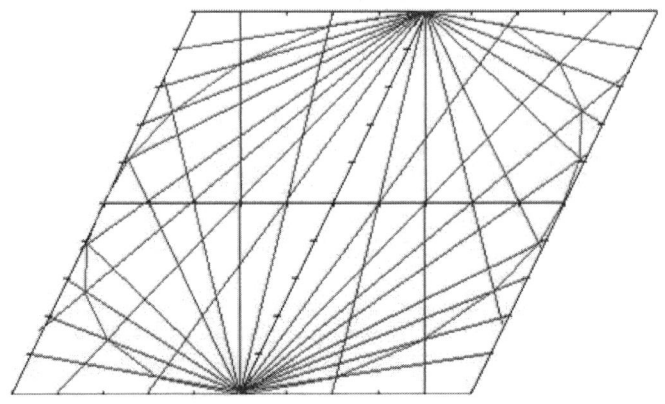

Figure 9-14: Constructing the parallelogram

This construction will work perfectly well if the parallelogram is a rectangle, so it will work to construct an ellipse if the major and minor axes are known. If the parallelogram is a square, the resulting ellipse is a circle. The intersecting lines are perpendicular, and the construction is the famous one of constructing a right angle inside a semicircle. The general construction here simply works by deforming the construction so an ellipse results.

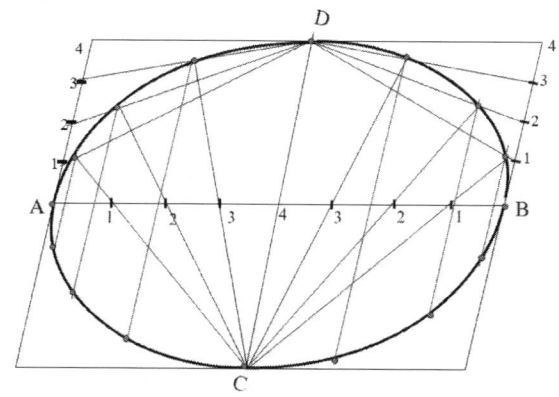

Figure 9-15: Constructing ellipse by oblong method

Solved examples

1. Drawing ellipse by pin-and-string method given the axis

To draw an ellipse by the pin-and-string method when the major axis, AB, and the minor axis, CD, are given, proceed as follows

Procedure:

a. With center C and radius equal to OA draw an arc intersecting the major axis at points F1 and F2, which are the foci.
b. Now fix the ends of a string at points F1 and F2, such that the length of the string is equal to AB.
c. For any point on the ellipse, such as point P or P, the sum of distances PF1 and PF2 (or of P'F1 and P'F2) remains equal to the constant length of the string.

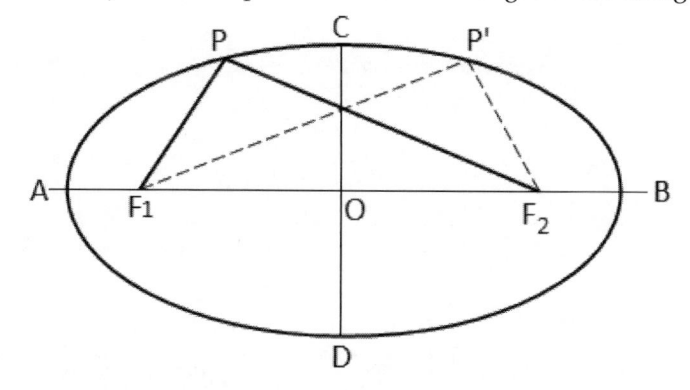

Figure 9-16: Pin-and-string method

Therefore, the ellipse is easily drawn by maintaining taut segments of the string, as a pencil (or other marking device) is used to draw the curve.

2. *To determine the major and minor axes, and the foci of a given ellipse*

Procedure:

The following processes are taken:

a. First draw two parallel lines such as m and n in Figure (9-17).
b. Draw AB, which bisects lines m and n.
c. Locate O, the midpoint of AB.
d. With O as center and radius OA, draw a circle to intersect the ellipse at points C and B.
e. Through O draw a line NN_1 parallel to line CA and draw line MM_1 perpendicular to CA.
f. Lines NN_1 and MM_1 are the minor and major axes, respectively.

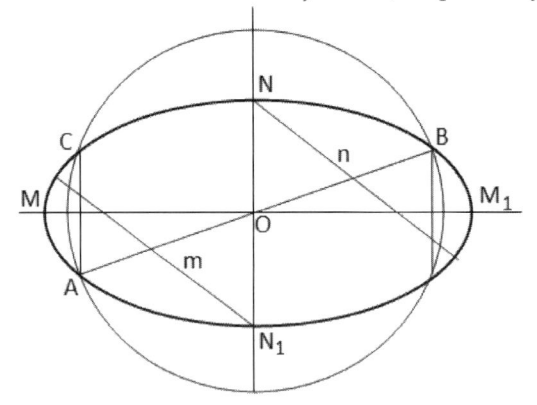

Figure 9-17: Determine the major and minor axes

To draw an ellipse by the concentric circle method, given the lengths of the major and minor axes.

Procedure:

a. Draw lines AB and CD as the given length of the major and minor axes, respectively.
b. Through point O, the center of the ellipse, draw radial lines such as m to intersect the concentric circles having radii OB and OC at points E, F, G, and K.

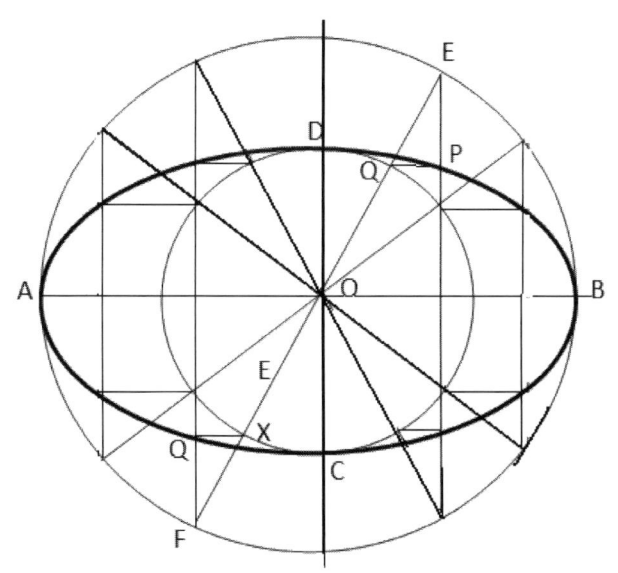

Figure 9-18: To draw an ellipse by the concentric circle method

c. Through points E and F draw vertical lines to intersect the horizontals drawn through G and K, in points P and Q, which are two points on the ellipse.
d. Repeat this construction for additional points and then draw a smooth curve through these points to form the ellipse.
e. The tangent, t, at point P passes through point R, which is the intersection of the tangent t' at point E of the major circle and the major axis extended (Figure 9-19).

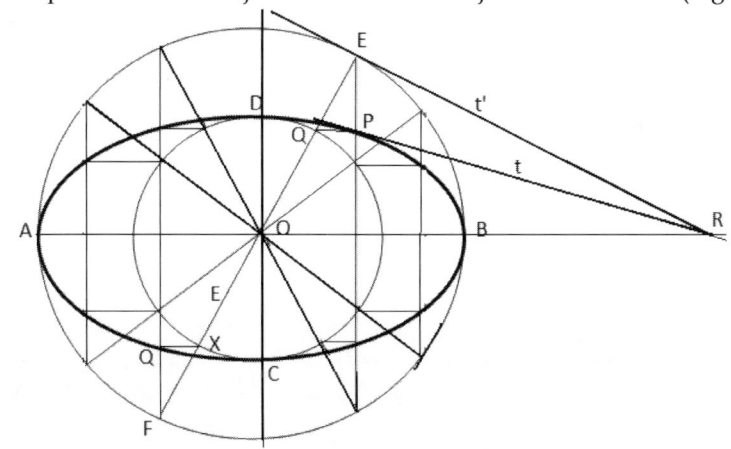

Figure 9-19: To draw an ellipse by the concentric circle method

To draw an ellipse by the use of circular arcs when the axes are given

To draw an ellipse by the use of circular arcs when the axes are given will result in a close approximation to a true ellipse.

Procedure:

1. Draw and mark the two given axes perpendicular to each other such that the minor axis CD bisect the major axis AB
2. Join points A and C.
3. Lay off distance CD equal to CE (where CE = OA-OC).

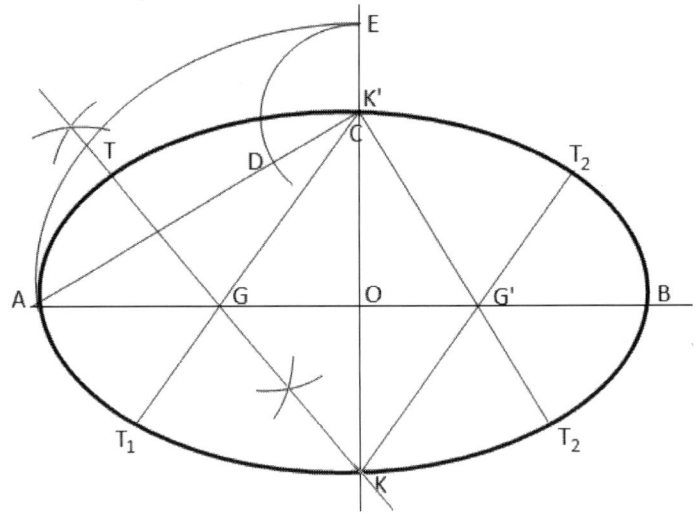

Figure 9-20: Drawing ellipse using circular arcs and given axes

4. Now draw the perpendicular bisector of AD and locate points G and K.
5. With G as center and radius GA, describe arc TAT_1.
6. With K as center and radius KT, describe arc TCT_2. Center G' and radius G'B are used to draw arc T_2BT_3; and center K' and radius K'T_3 are used to draw arc T_3T_1.

To inscribe an ellipse into a given rectangle

Procedure:

1. Divide OA into a number of equal parts (four are shown), and then divide AE into the same number of equal parts.
2. Now draw rays D-1, D-2, etc., to intersect the corresponding rays C-1, C-2, etc., in points which lie on the ellipse.
3. The construction shown may be repeated for the other quarters of the rectangle in order to obtain additional points on the ellipse.

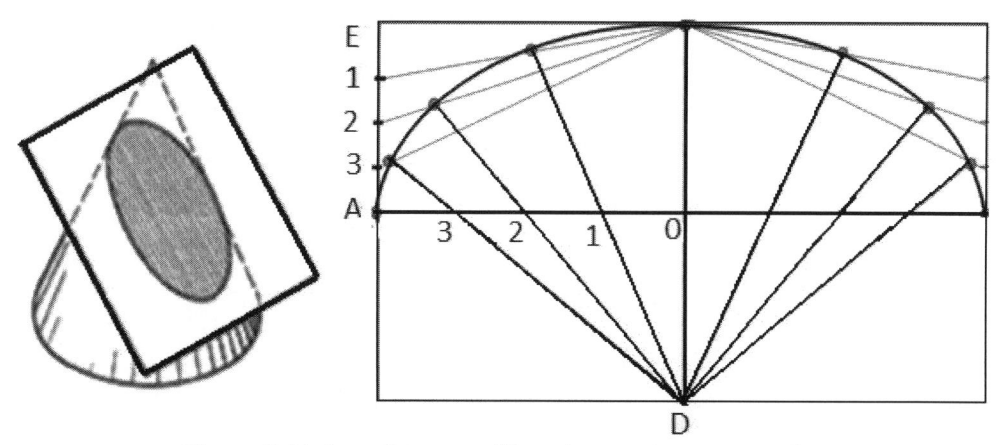

Figure 9-21: Inscribing an ellipse into a given rectangle

The pictorial shows a right circular cone intersected by an inclined plane that cuts all the elements of the cone. The intersection is an ellipse.

9.2.2 Hyperbola and its construction

The hyperbola may be defined as the locus of a point which moves such that the ratio of its distances from the focus and the directrix is constant and greater than 1. It is also explained as the locus of all coplanar points the differences of whose distances from two fixed points (foci) are a constant. A *hyperbola* is therefore a set of all points P(x, y) such that the *difference* of the distances between (x, y) and two distinct points is a constant. The fixed points are called the foci of the hyperbola.

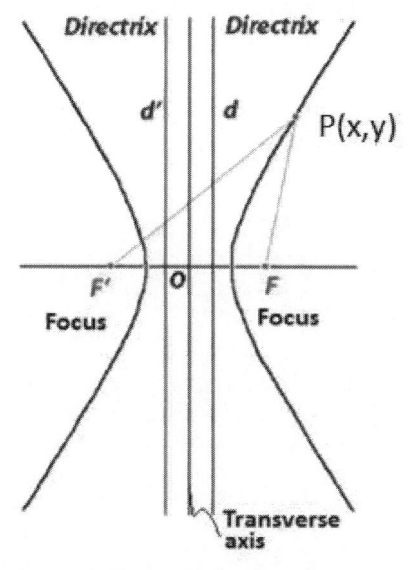

Figure 9-22: Hyperbola on transverse axis

The graph of a hyperbola has two parts, called branches. Each part resembles a parabola but is a slightly different shape. A hyperbola has two vertices that lie on an axis of symmetry called the transverse axis. For the hyperbolas studied here, the transverse axis is either horizontal or vertical.

Basic geometrical properties employed in constructing a hyperbola

1. Equation of a hyperbola with center at the origin

Let *a* and *b* represent positive real numbers.

Horizontal transverse axis: The standard form of an equation of a hyperbola with a *horizontal transverse axis* and center at the origin is given by

$$\frac{x^2}{a^2} - \frac{y^2}{b^2} = 1$$

Note: The *x*-term is positive. The branches of the hyperbola open left and right.

Vertical transverse axis: The standard form of an equation of a hyperbola with a *vertical transverse axis* and center at the origin is given by

$$\frac{x^2}{b^2} - \frac{y^2}{a^2} = 1$$

Note: The *y*-term is positive. The branches of the hyperbola open up and down. In the standard forms of an equation of a hyperbola, the right side must equal 1.

Constructing hyperbola

Constructing a hyperbola using the line of eccentricity

The eccentricity of the hyperbola we are dealing with is 4/3; therefore the line of eccentricity will be greater than 45.°

Principle: For every point on the curve the distance to the focal point over the distance to the directrix is in a ratio of 4/3.

$$\frac{PF}{PD} = \frac{4}{3}$$

Procedure: The procedure for constructing the hyperbola is the exact same as that described for the ellipse.

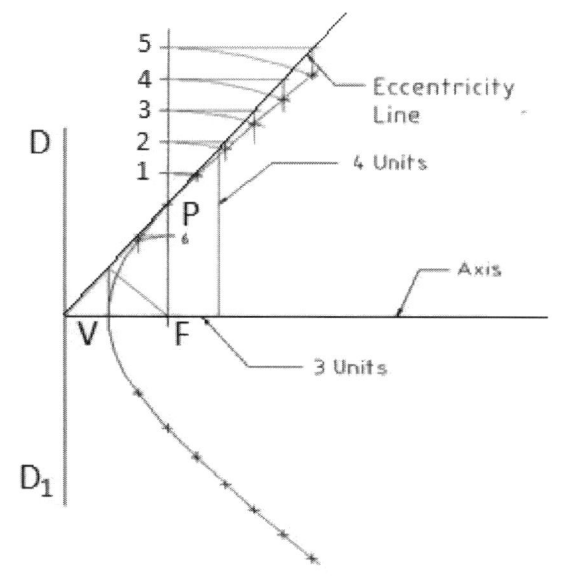

Figure 9-23: Constructing hyperbola by eccentricity method

Constructing a hyperbola given the foci and constant distance

To construct a hyperbola when the foci, F_1 and F_2, and the constant distance, AB, are given

Procedure:

c. With F1 as center and a radius greater than FIB, an arc is drawn.
d. With F2 as center and a radius which is equal to the difference between the first radius and length AB, an arc is drawn to intersect the first arc in points P and Q, which are two points on the hyperbola. It is clearly seen that $F_1P-F_2P = AB$ and that $F_1Q-F_2Q = AB$.
e. Additional points may be found in a similar manner.
f. Draw the smooth curve passing through the points as the hyperbola.

Note: It should be noted that the curve has two branches which are symmetrical with respect to the axes. The asymptotes pass through the center O and are tangent to the curve at infinity. They are located by joining point O the center of the hyperbola with points K and K'. These points are found by locating the intersections of the verticals through points A and B with the circle of radius OFT. The tangent, t, at point P bisects the angle F_2PF_1.

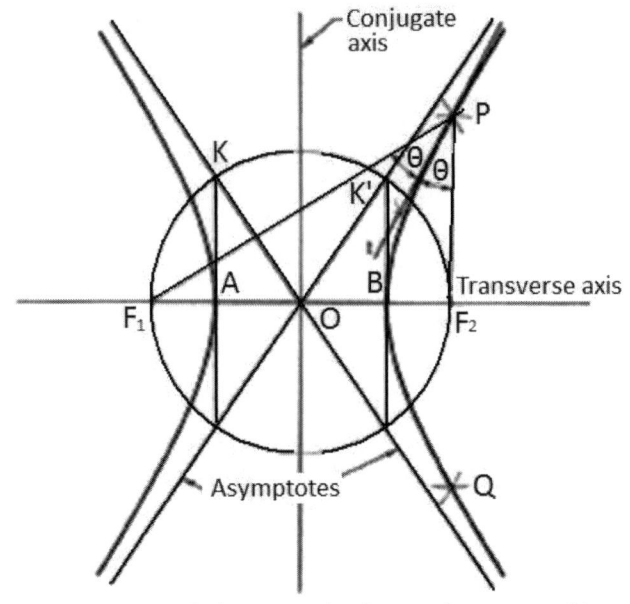

Figure 9-24: Hyperbola given the foci and constant distance

Constructing a rectangular hyperbola

To construct a rectangular hyperbola (asymptotes are at right angles), given the asymptotes m and n, and one point P on the curve.

Procedure:

1. Draw lines k and t through point P, respectively parallel to n and m.
2. Select any point Q on line k and then draw line OQ.
3. Locate point R, the intersection of OQ and t.

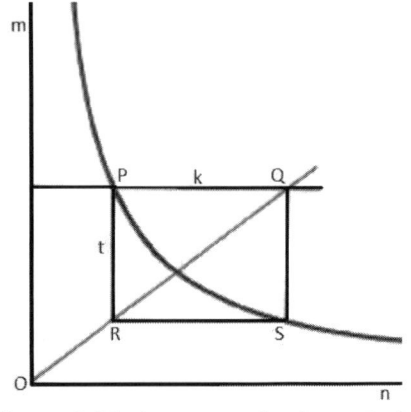

Figure 9-25: A rectangular hyperbola

4. Draw a horizontal line through point R and a vertical line through point Q.
5. The intersection of these lines is point S, a point on the hyperbola.
6. In a similar manner additional points are located.

Constructing a hyperbola, given the transverse axis and a point

To construct a hyperbola, given the transverse axis AB, and a point P on the curve, proceed as follows.

Procedure

1. First construct the rectangle PCDE.
2. Divide side EP into a number of equal parts (four are shown) and the right half of side CP into the same number of equal parts.
3. Find the intersection of rays A-1, A-2, etc., with the corresponding rays B-1, B-2, etc.

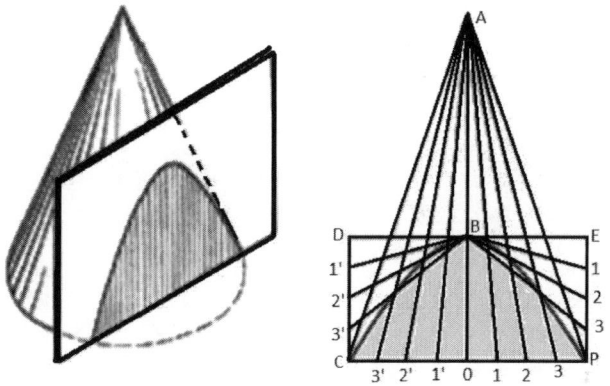

Figure 9-26: Hyperbola, given the transverse axis and a point

4. Repeat the procedure for the left half of the rectangle.
5. The smooth curve which passes through the points thus located is one branch of the hyperbola. The other branch may be determined in a similar manner. The pictorial shows a right circular cone intersected by a plane parallel to the axis of the cone. The intersection is a hyperbola (one branch shown).

9.2.3 Parabola and its construction

A parabola is defined as the locus of a point which moves such that its distance from a fixed point, the focus, and a fixed straight line, the directrix, are always equal. Parabola is further defined by a set of points in a plane that are equidistant from a fixed line (called the directrix) and a fixed point (called the focus) not on the directrix.

Parabolas have numerous real-world applications. For example, a reflecting telescope has a mirror with the cross section in the shape of a parabola. A parabolic mirror has the property that incoming rays of light are reflected from the surface of the mirror to the focus.

Basic geometrical properties employed in constructing a parabola

1. The *axis of symmetry* of the parabola is a line that passes through the vertex and is perpendicular to the directrix
2. *Equation of a parabola with vertical axis of symmetry*

    The standard form of the equation of a parabola with vertex and vertical axis of symmetry is

    $$y = a(x - h)^2 + k$$

    Where a≠ 0; If a>0, then the parabola opens upward; and if a<0, the parabola opens downward.

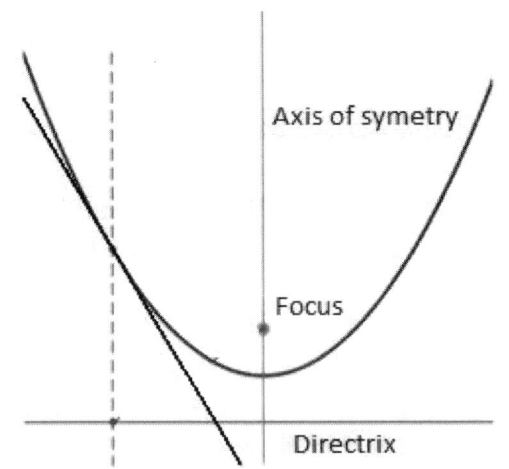

    Figure 9-27: Parabola on vertical axis of symmetry

    The axis of symmetry is given by $x = h$.

3. *Equation of a parabola with horizontal axis of symmetry*

    The standard form of the equation of a parabola with vertex and vertical axis of symmetry is

    $$x = a(y - h)^2 + k$$

Where a≠ 0; If a>0, then the parabola opens to the right; and if a<0, the parabola opens to the left.

The axis of symmetry is given by $x = h$.

4. The graph of a parabola expressed by the equation $y = ax^2 + bx + c$ will open upward if a>0 and open downward if a<0.
5. A parabola can also open to the left or right. In such a case, the "roles" of x and y are essentially interchanged in the equation.

Thus, the graph of $x = ay^2 + by + c$ opens to the right if a>0 and (Figure 9-28 left) and to the left if a<0 (Figure 9-28 right).

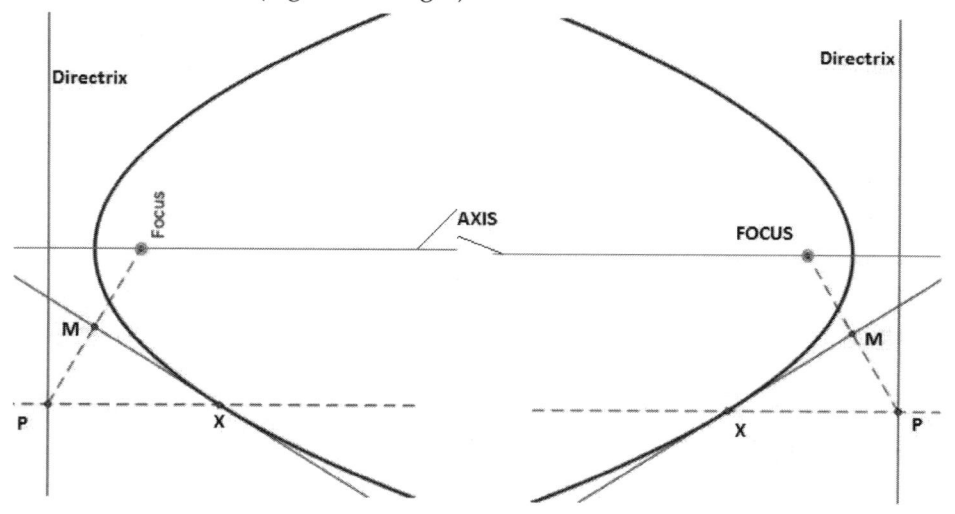

Figure 9-28: Parabola on horizontal axis of symmetry

Constructing hyperbola by eccentricity method

In constructing a parabola, a general procedure to be taken includes:

1. Construct a line that will be the directrix
2. Construct a point not on the line to be the focus
3. Construct a free point P on the directrix
4. Construct a line through P that is perpendicular to the directrix
5. Connect P and the focus.
6. Construct the perpendicular bisector of the line connecting P and the focus through the midpoint M.
7. Mark the intersection of the perpendicular bisector and the line perpendicular to the directrix through P as X.

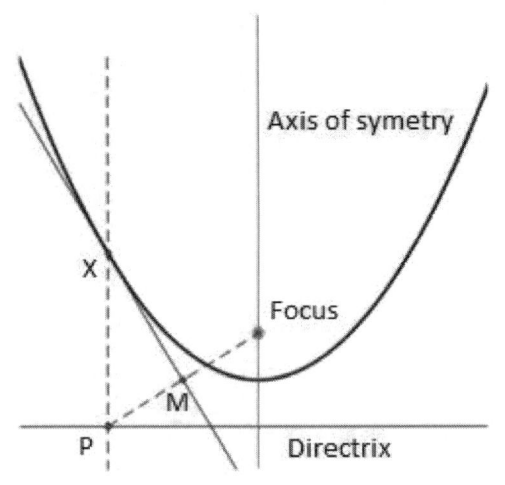

Figure 9-29: Constructing hyperbola by eccentricity method

8. -X will trace the parabola as P moves along the directrix
9. Line XM is the line tangent to the parabola

Constructing a parabola using the line of eccentricity

The eccentricity of a parabola is always 1; therefore the line of eccentricity is always at 45° to the axis.

Principle: From each point on the curve the distance to the focal point is the same as the distance to the directrix.

$$\frac{PF}{PD} = 1$$

Procedure

The procedure as can be seen is the exact same as for the ellipse. When getting the vertex, a line is projected at 45° to the line of eccentricity, which for the parabola is at 45° to the axis, and then dropped perpendicular to the axis; therefore the vertex will be half way between the directrix and the focus. Like the ellipse the other vertex is found by projecting another 45° line to meet the line of eccentricity, which as just stated is also at 45,° therefore they will only meet at infinity, so the other vertex is at infinity.

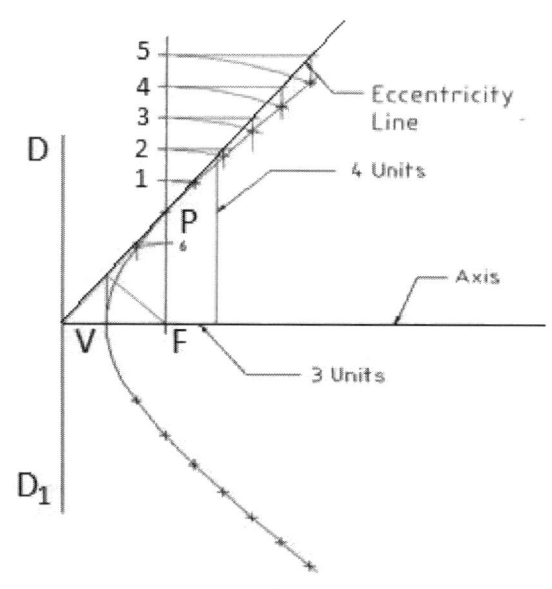

Figure 9-30: Construction of parabola by eccentricity

Construction of parabola

To locate points on a parabola when the focus, F, and the directrix, d, are given.

Procedure

Points such as 1 and 2 are determined by locating the intersection of line s (any line parallel to the directrix) and an arc having center F and a radius equal to the distance between the parallel lines d and s. Now it is quite apparent that points 1 and 2 are the same distance from F, the focus, as they are from d, the directrix. The tangent, t, at point P bisects the angle KPF.

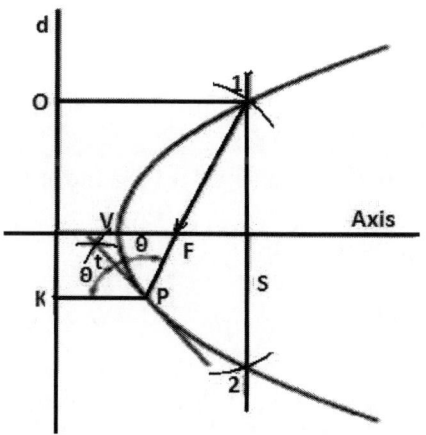

Figure 9-31: Locating points on a parabola

To determine the axis, focus, and directrix of a given parabola

Procedure

The axis is located in the following manner.

1. Draw two parallel chords such as m and n.
2. The line t joining the midpoints of these chords is parallel to the axis.
3. Introduce a line such as s perpendicular to line t.
4. The required axis is the perpendicular bisector of line s.

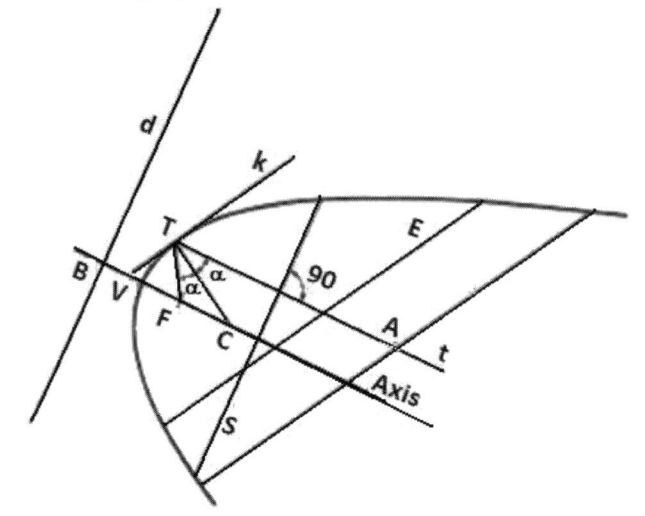

Figure 9-32: Determining the axis, focus, and directrix of a given parabola

The focus, F, is located by making angle FTC equal to angle CTA. Point C is the intersection of the axis with the perpendicular to tangent line k at point T. The directrix, d, is perpendicular to the axis and at a distance from V equal to VF, that is, VB = VF.

Constructing parabola given the axis, vertex, and a point through which the parabola passes

To construct a parabola, given the axis, vertex V, and a point P through which the parabola passes.

Procedure

2. First draw rectangle PABC.
3. Divide CP and CV into the same number of equal parts.

4. Introduce lines parallel to the axis and passing through points 1, 2, and 3 on side VC.
5. Draw rays V-1, V-2, and V-3.
6. Finally locate the points in which the parallels intersect the corresponding rays, i.e., the parallel through point 1 intersects ray V-1, etc. The curve through the points thus located is the parabola.

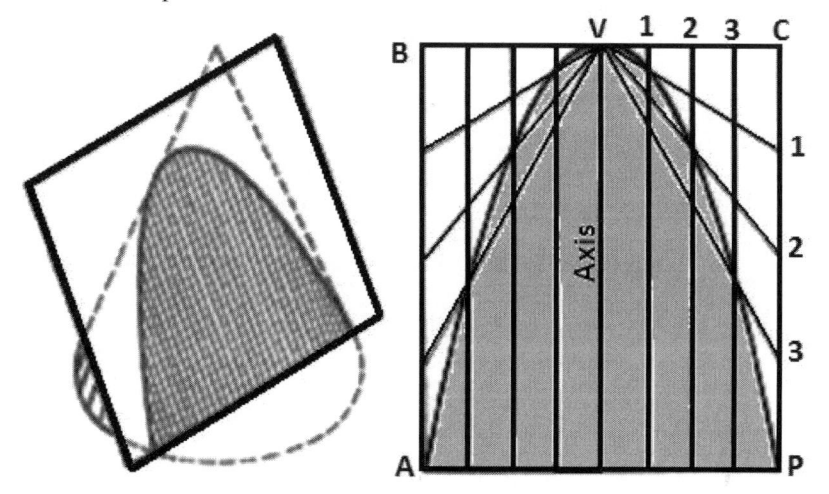

Figure 9-33: Constructing parabola given the axis, vertex, and a point

The pictorial shows a right circular cone intersected by a plane parallel to an element of the cone. The intersection is a parabola.

Exercise

2. Define the key terms.
    a. Conic sections b. Ellipse c. Hyperbola d. Transverse axis of a hyperbola e. Vertex
3. An arch for a tunnel is in the shape of a semi-ellipse. The distance between vertices is 120 ft, and the height to the top of the arch is 50 ft. find the height of the arch 10 ft from the center. Round to the nearest foot.

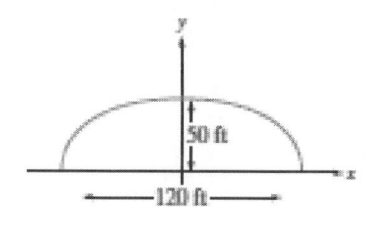

4. A bridge over a gorge is supported by an arch in the shape of a semi-ellipse. The length of the bridge is 400 ft, and the height is 100 ft. find the height of the arch 50 ft from the center. Round to the nearest foot.

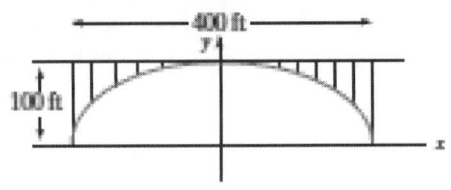

Chapter 10

Solid Sections and Development

10.1 Introduction

The basic methods of representing an object by projecting them in views have been discussed in previous chapters. These methods allow us to see the external features of such object only. Often times it is necessary to view the internal features of the object, this could only be accomplished by slicing through the object and producing a *sectional* or *section view*

Sections and sectional views

Sections and sectional views are used to show hidden detail more clearly. They are created by using cutting planes to cut the object. A section is a view of no thickness and shows the outline of the object at the cutting plane. Visible outlines beyond the cutting plane are not drawn. A sectional view, displays the outline of the cutting plane and all visible outlines which can be seen beyond the cutting plane. The cutting planes (section lines) are always a phantom line type.

The diagram below shows a sectional view, and how a cutting plane works.

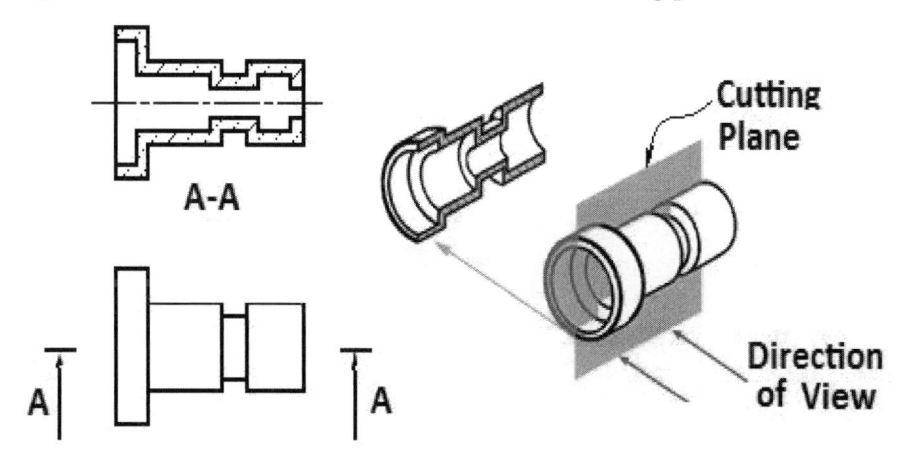

Figure 10-1: A sectional view

Cross-sectional views

A cross-sectional view portrays a cut-away portion of an object and it is another way of showing hidden components in a device. Imagine a plane that cuts vertically through the center of the pillow block as shown in Figure 10-2. Then imagine removing the material (part in shade on the left hand side) from the front of this plane, as shown in Figure 10-2, the remaining cut part is shown in the figure to the right hand side.

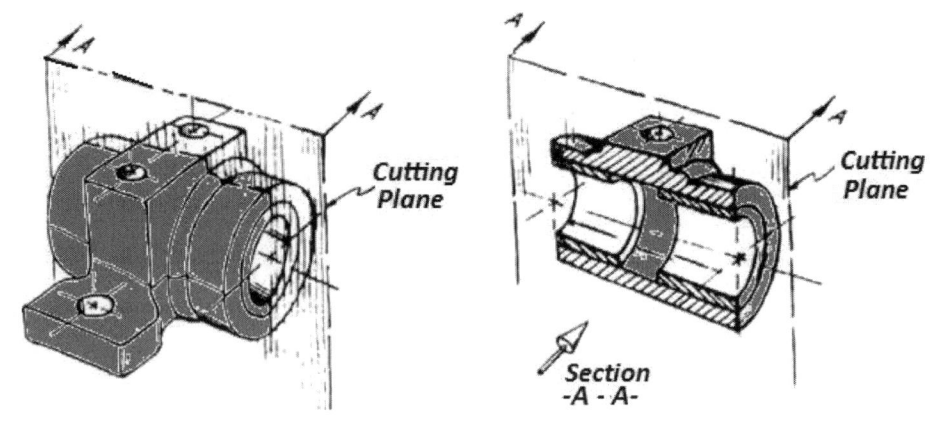

Figure 10-2: Pillow block

Figure 10-3 show how the remaining rear section in the direction of arrow (Section AA) would look. Diagonal lines (cross-hatches) show regions where the material has been cut by the cutting plane. It is conventional that section views are always placed behind the arrows.

Section -A-A-

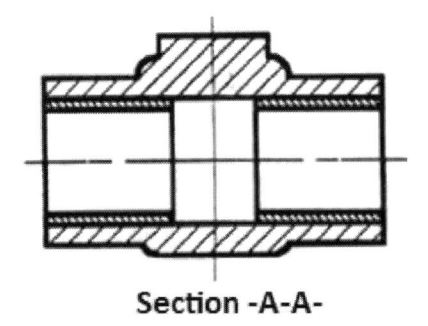

Figure 10-3: Section A-A

This cross-sectional view (section A-A, in Figure 10-3), one that is orthogonal to the viewing direction, shows the relationships of lengths and diameters better. These drawings are easier to make than isometric drawings. Seasoned engineers can interpret orthogonal drawings without needing an isometric drawing, but this takes a bit of practice.

The top "outside" view of an orthogonal (perpendicular) bearing is shown in Figure 10-4. It is. Notice the direction of the arrows for the "A-A" cutting plane.

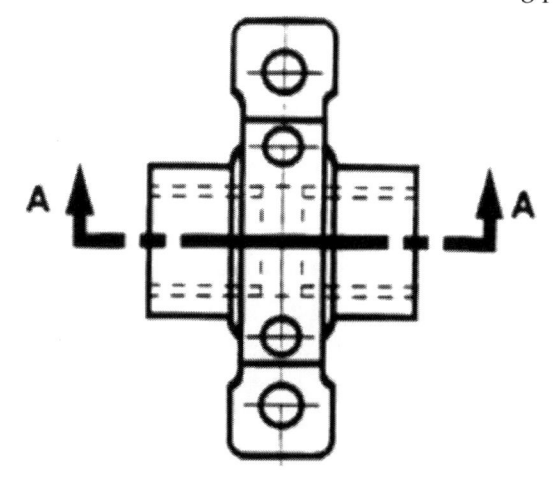

Figure 10-4: The top 'outside' view of the bearing

Benefits of sectional views

Sectional views are extremely useful in minimizing the number of projected views. They provide clear and unambiguous representation of internal features. Section views can reduced the number of views of many axis-symmetric parts to a single view. For example the drawing to the left hand side of the Figure 10-5 below is unnecessary because the sectional view at the right gives a detailed feature of the drawing.

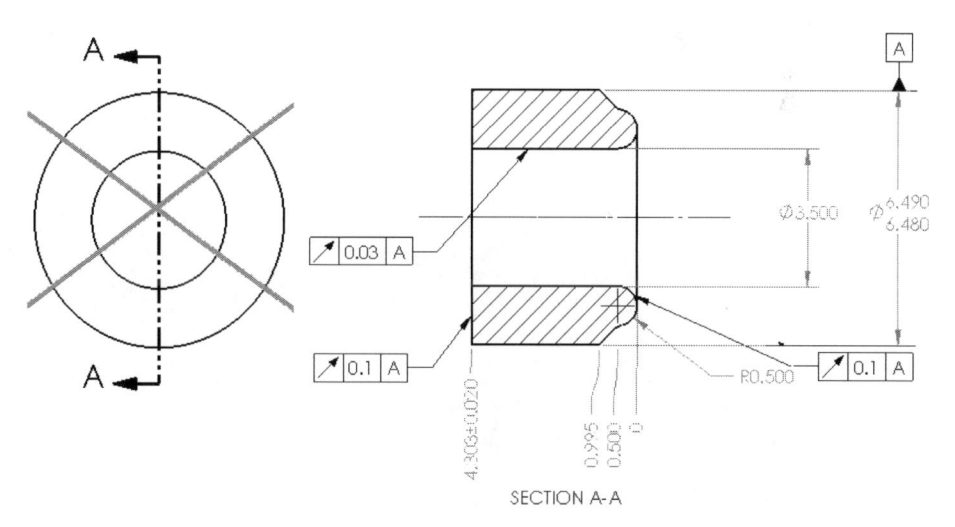

Figure 10-5: Sectioning axis-symmetry object

10.2 Types of sections

There are different types of sectional views that are suitable for sectional drawings depending on the required views. Below are typical examples.

1. *Sectional view in a single plane*

The example below shows a simple single plane sectional view where object is cut in half by the cutting plane. The cutting plane is indicated on a drawing using the line style used for center lines, but with a thick line indicating the end of lines and any change in the direction of the cutting plane. The direction of the view is indicated by arrows with a reference letter. The example below shows a sectional view of the cutting plane A - A.

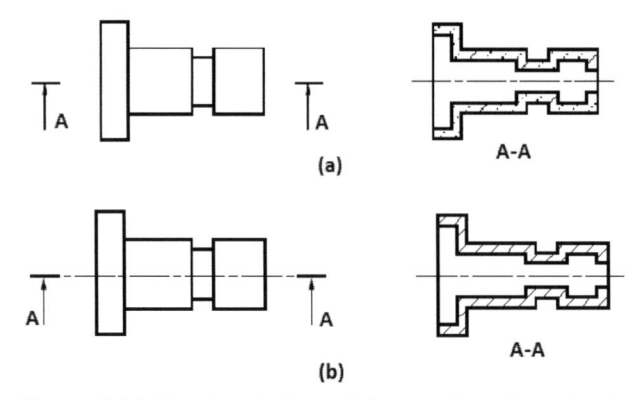

Figure 10-6: Sectional view of the cutting plane A - A

v. *Sectional view in two planes*

It is possible for the cutting plane to change directions, to minimize on the number of sectional views required to capture the necessary detail. The example below shows a pipe being cut by two parallel planes. The sketch shows where the object is cut.

Figure 10-7: Sectional view in two planes

The section drawing of the cut section is shown in two dimensional views represented by the orthographic views shown.

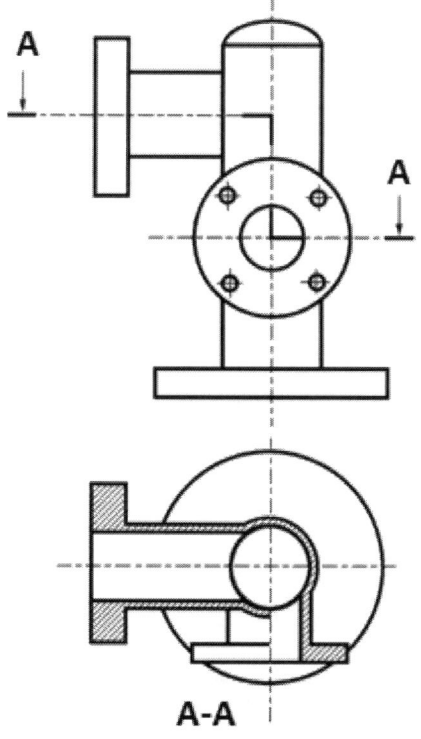

Figure 10-8: Sectional view in two planes

vi. *Half sectional views*

Half sections are commonly used to show both the internal and outside view of symmetrical objects. A half-section is a view of an object showing one-half of the view in section, as in Figure 10-9 and 10-10.

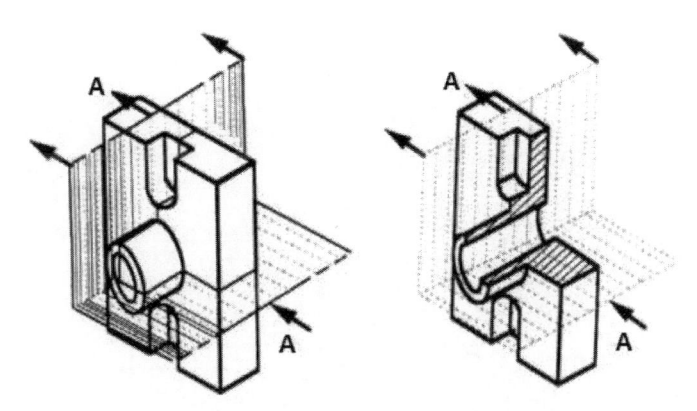

Figure 10-9: Full and sectional views

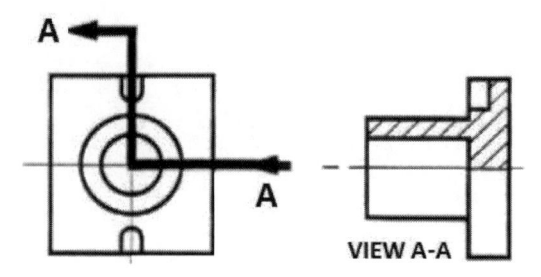

Figure 10-10: Front view and half section

vii. Part sectional views

It is common practice to section a part of an object when only small areas are needed to be sectioned to indicate important details. The example in Figure 9-10 shows a part sectional view to indicate a through-hole in a plate. Notice that the line indicating the end of the section is a thin continuous line.

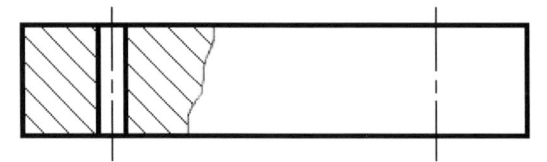

Figure 10-11: Part sectional view

viii. Broken-out sectional view

Broken-out Section views are essentially partial section views without the section arrow. Often times they are used to expose a feature of interest while eliminating the need to create another view.

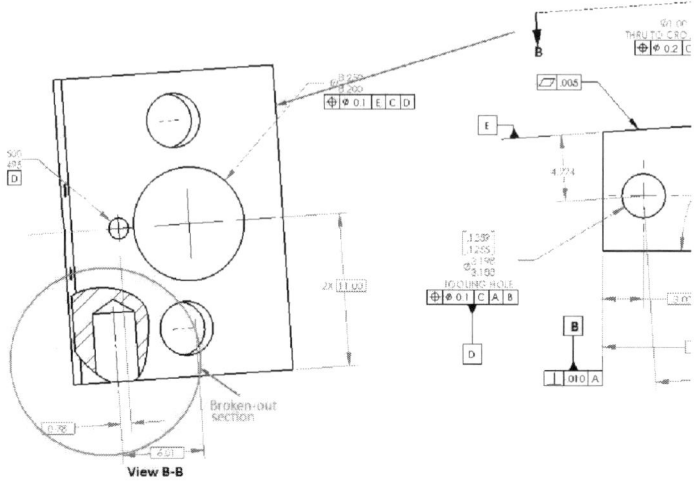

Figure 10-12: Broken-out sectional view

248 | Page

ix. *Partial sectional view*

Partial views are removed views and are established in a similar manner as section views, that is they require view arrows to establish viewing direction. However, they do not have to section an entire object; rather can simply display a partial view of a projection at a larger scale if desired.

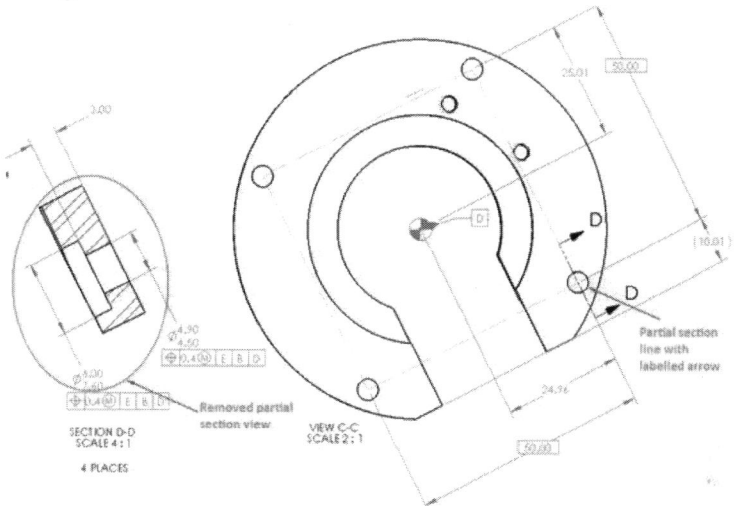

Figure 10-13: Partial sectional view

x. *Cropped view*

Cropped views reduce the size of a view such that only necessary information is displayed. Cropped views also maximize the sheet area by reducing view size.

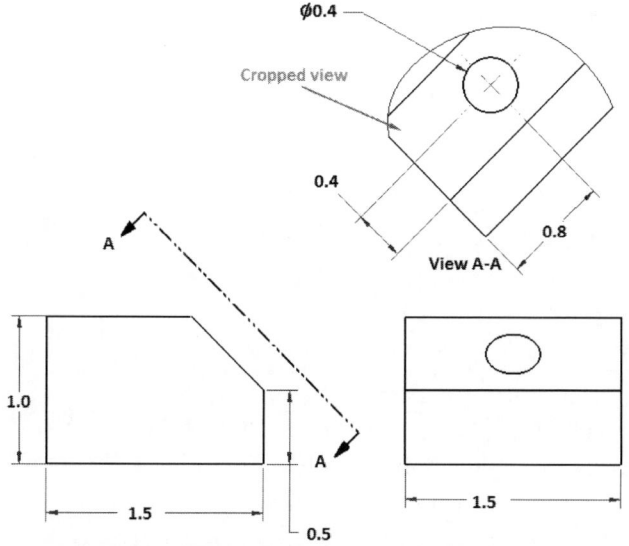

Figure 10-14: Drawing showing cropped view

249 | Page

Sectioning objects with holes, ribs etc

The cross-section on the right of Figure 9-15 is technically correct. However, the convention in a drawing is to show the view on the left as the preferred method for sectioning this type of object.

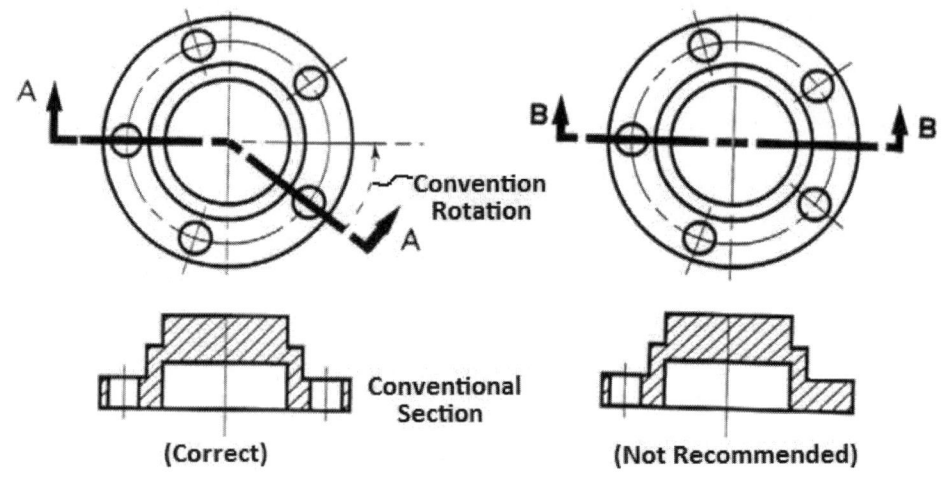

Figure 10-15: Cross-section conventions

There are many times when the interior details of an object cannot be seen from the outside (such as in Figure 10-16). We can get around this by pretending to cut the object on a plane and showing the "sectional view". The sectional view is applicable to objects like engine blocks, where the interior details are intricate and would be very difficult to understand through the use of "hidden" lines (hidden lines are, by convention, dotted) on an orthographic or isometric drawing.

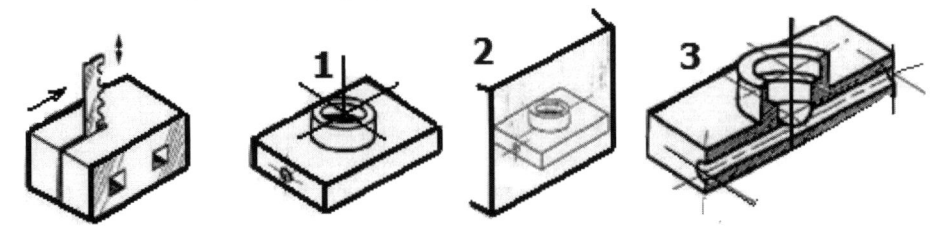

Figure 10-16: Sectioning steps

10.3 Hatching

On sections and sectional views solid area should be hatched to indicate this fact. Hatching is drawn with a thin continuous line, equally spaced (preferably about 4mm apart, though never less than 1mm) and preferably at an angle of 45 degrees.

The diagonal lines on the section drawing used to indicate the area that has been theoretically cut is called hatching. These lines are called section lining or cross-hatching. The lines are thin and are usually drawn at a 45-degree angle to the major outline of the object. The spacing between lines should be uniform.

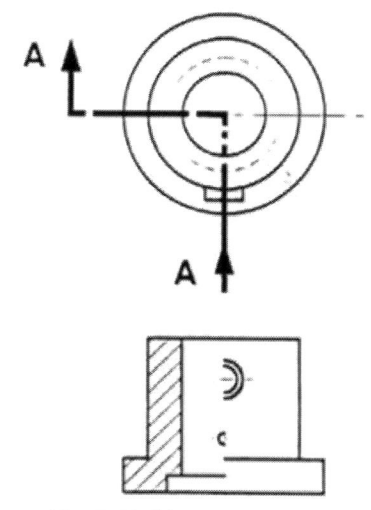

Figure 10-17: Half section without hidden lines

A second, but rarer, use of cross-hatching is to indicate the material of the object. One form of cross-hatching may be used for cast iron, another for bronze, and so forth. More usually, the type of material is indicated elsewhere on the drawing, making the use of different types of cross-hatching unnecessary.

Usually hidden (dotted) lines are not used on the cross-sections unless they are needed for dimensioning purposes. Also, some hidden lines on the non-sectioned part of the drawings are not needed since they become redundant information and may clutter the drawing.

Hatching a single object

When you are hatching an object, but the objects have areas that are separated, all areas of the object should be hatched in the same direction and with the same spacing (Figure 10-18).

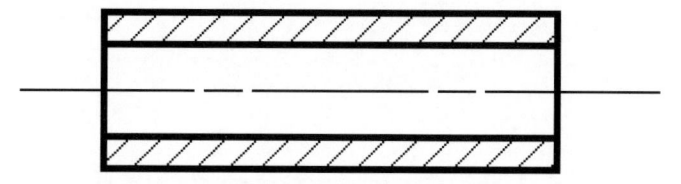

Figure 10-18: Object hatched in the same direction

Hatching adjacent objects

When hatching assembled parts, the direction of the hatching should ideally be reversed on adjacent parts. If more than two parts are adjacent, then the hatching should be staggered to emphasize the fact that these parts are separate Figure 10-19.

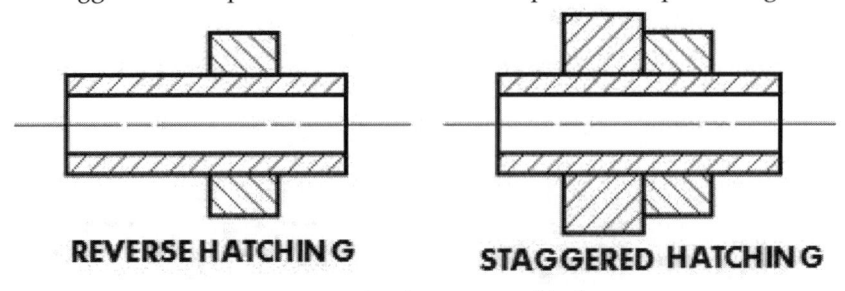

REVERSE HATCHING   **STAGGERED HATCHING**

Figure 10-19: Hatching assembled parts

Hatching thin materials

Sometimes, it is difficult to hatch very thin sections. To emphasize solid wall, the walls can be filled in. This should only be used when the wall thickness size is less than 1mm

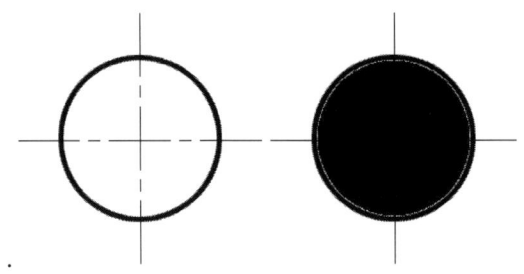

Figure 10-20: Hatching thin sections

Hatching large area

When hatching large areas in order to aid readability, the hatching can be limited to the area near the edges of the part.

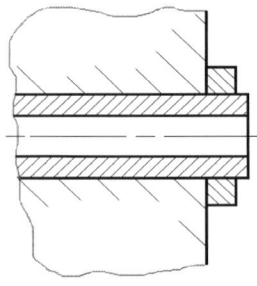

Figure 10-21: Hatching large areas

Drawing conventions for threaded parts

Threads are drawn with thin lines as shown in this illustration. When drawn from end-on, a threaded section is indicated by a broken circle drawn using a thin line.

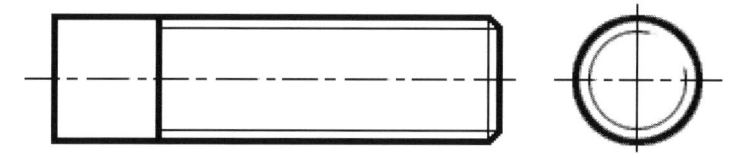

Figure 10-22: Convention for threaded parts

Frequently a threaded section will need to be shown inside a part. The two illustrations to the left demonstrate two methods of drawing a threaded section. Note the conventions. The hidden detail is drawn as a thin dashed line. The sectional view uses both thick and thin line with the hatching carrying on to the very edges of the object.

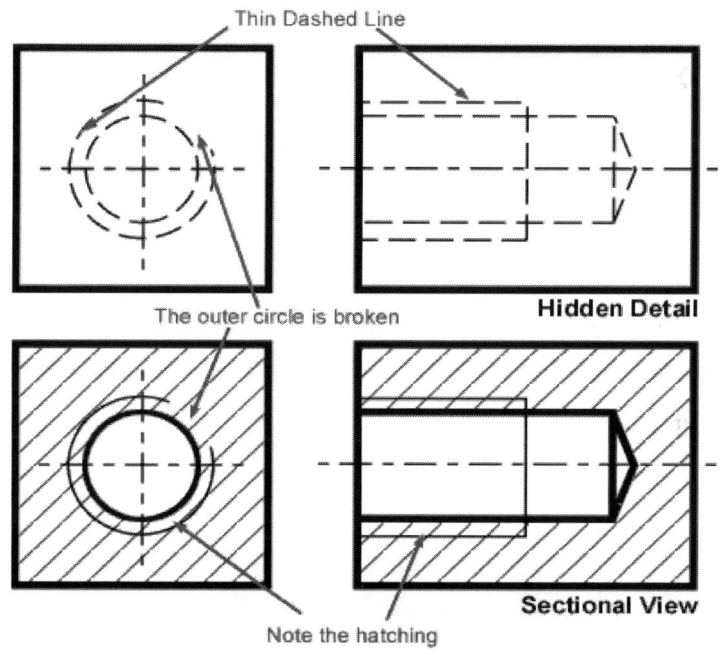

Figure 10-23: Sectional view of threaded parts

10.4 Sections of solid

A section is formed by the intersections of a plane and a solid. The plane may be inclined, vertical or oblique and is known as a cutting plane. Sections are used to show the construction of buildings or objects and of any internal arrangement.

True surface of rectangular prism

An example shows a rectangular prison cut by a vertical plane MN as shown. To determine the true shape of the section;

1. Draw the plan and elevation
2. Project parallel to the cutting plane MN a new XY and draw the new elevation which gives the true shape of the section MN.

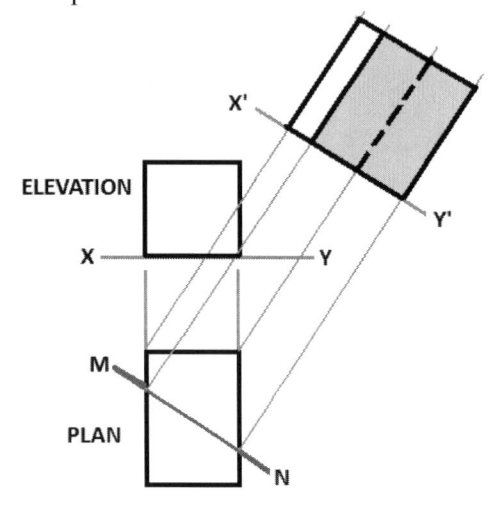

Figure 10-24: True shape of section

In a similar manner a situation where the rectangular prison represents a chimney stack intersecting a roof surface, represented by the inclined plane MN. Projections, from this plane give the true shape of the sections made by the plane.

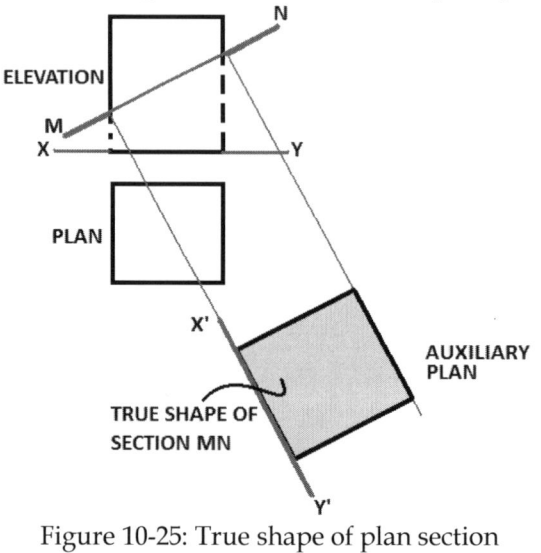

Figure 10-25: True shape of plan section

True surface of intersecting rectangular pyramid

The plan and elevation of a pyramidal roof intersecting a main pitched roof are shown in Figure 5-3.

1. Draw the elevation of the pyramidal roof and project the plan.
2. Project points MN in the elevation to the corresponding hips in the plane at A and B.
3. Complete the plan joining the straight lines.
4. Project the auxiliary plan from the elevations at *MN* making *A'B'* equal to *AB* and *C'D'* equal to *CD* to give the required shape of the section.

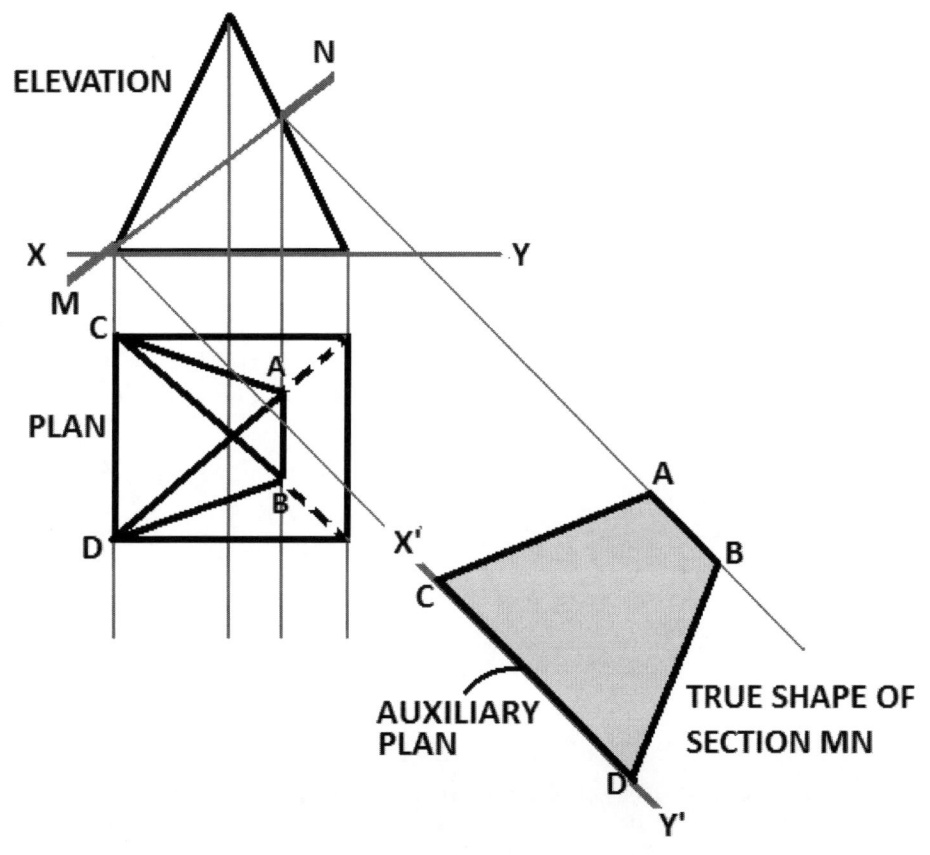

Figure 10-26: True shape of plan projected

Auxiliary surface of truncated cone

The plan and elevation of a cone resting on the H.P is shown in Figure 10-27. To draw a new or auxiliary plan of the truncated cone on the new *X'Y'* plane, proceed as follows:

1. Draw the plan and section
2. On a new XY line parallel to the cutting plane project the base of the cone from elevation to give the minor axis.
3. The major axis is equal to the diameter of the cone at its base.
4. Project the cutting plane MN to the auxiliary pan to give the major axis, the minor axis is taken from the plan of the section.

The two ellipses in the projected plan of the truncated cone have these corresponding axes reserved.

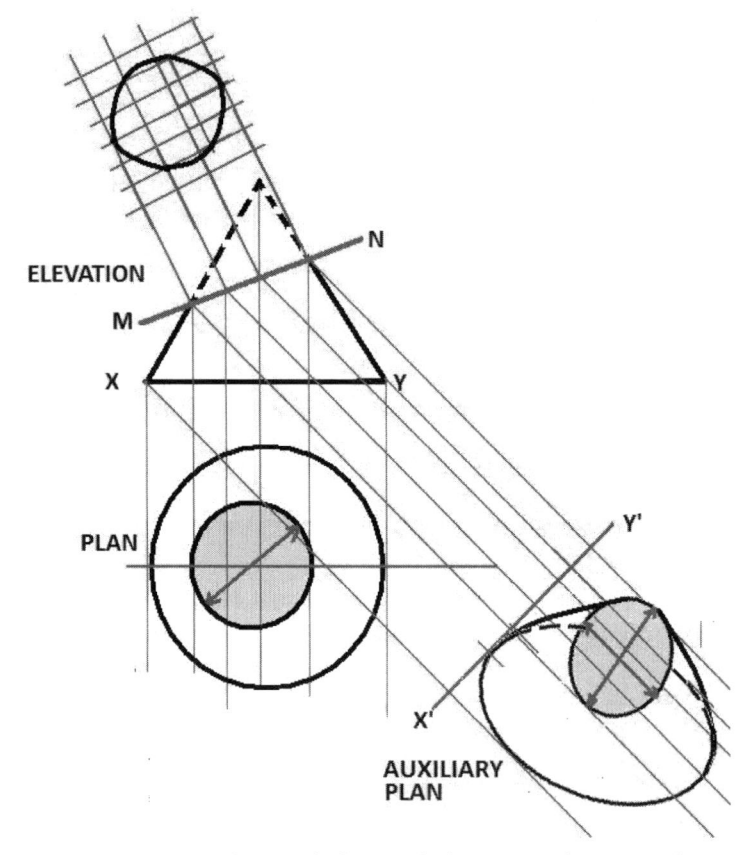

Figure 10-27: True shape of plan and elevation of cone surface

Sectional view and true shape of cone

A cone base 75 mm diameter and axis 100 mm long, has its base on the HP. A section plane parallel to one of the end generators and perpendicular to the FP cuts the cone intersecting the axis at a point 75 mm from the base. Draw the sectional Top View and the true shape of the section

1. The section plane is parallel to one of the end generators and perpendicular to the frontal plane
2. It is therefore drawn in the Front View
3. It cuts the axis at a point 75 mm from the base as shown in Front View

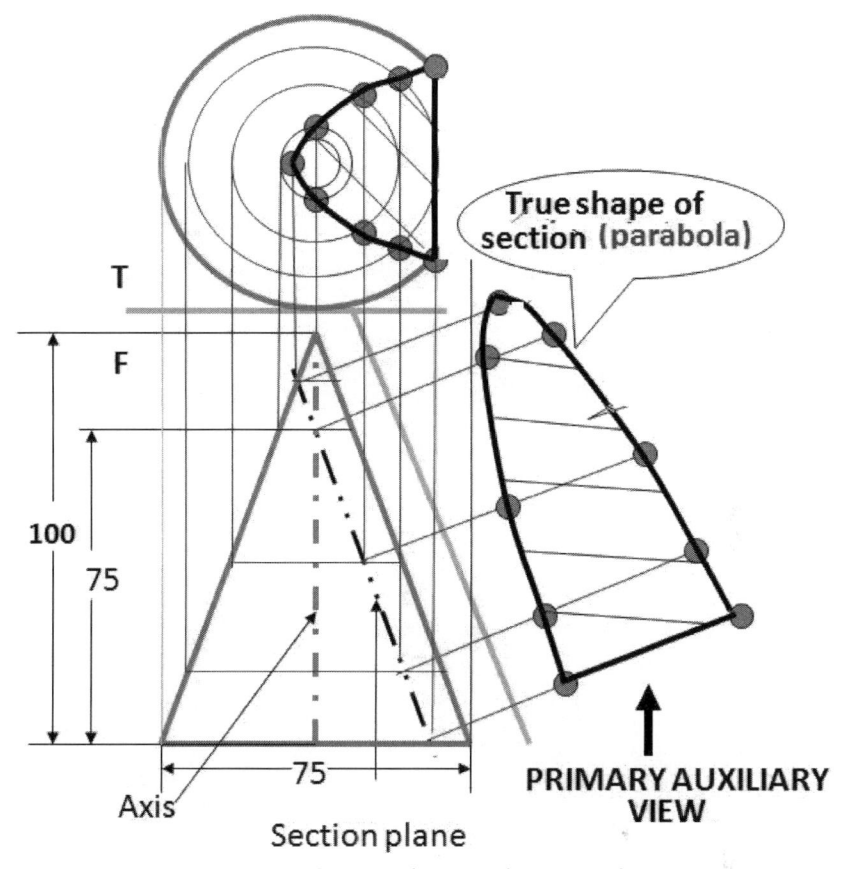

Figure 10-28: Sectional view of cone surface

4. Draw horizontal circles around the cone surface with center coinciding with the axis in the TV
5. Project corresponding points of intersection of the circles with the section plane in FV to the TV
6. Join these points to get the section face
7. For true shape of the section, draw an auxiliary view with reference line parallel to the section plane

True surface of pentagonal pyramid

To find the true surface of a pentagonal pyramid of 50 mm base side and height 100 mm resting on its base on the ground with axis parallel to frontal plane and perpendicular to the top plane. One of the sides of the base is closer and parallel to the frontal plane. A vertical section plane cuts the pyramid at a distance of 15 mm from the axis with section plane making an angle of 50° with FP. Draw the remaining part of the pyramid and the true shape of the cut section.

Solution

The figure is represented in Figure 10-29 below. The cutting plane is perpendicular to the top view; therefore the section line is drawn from the top view.

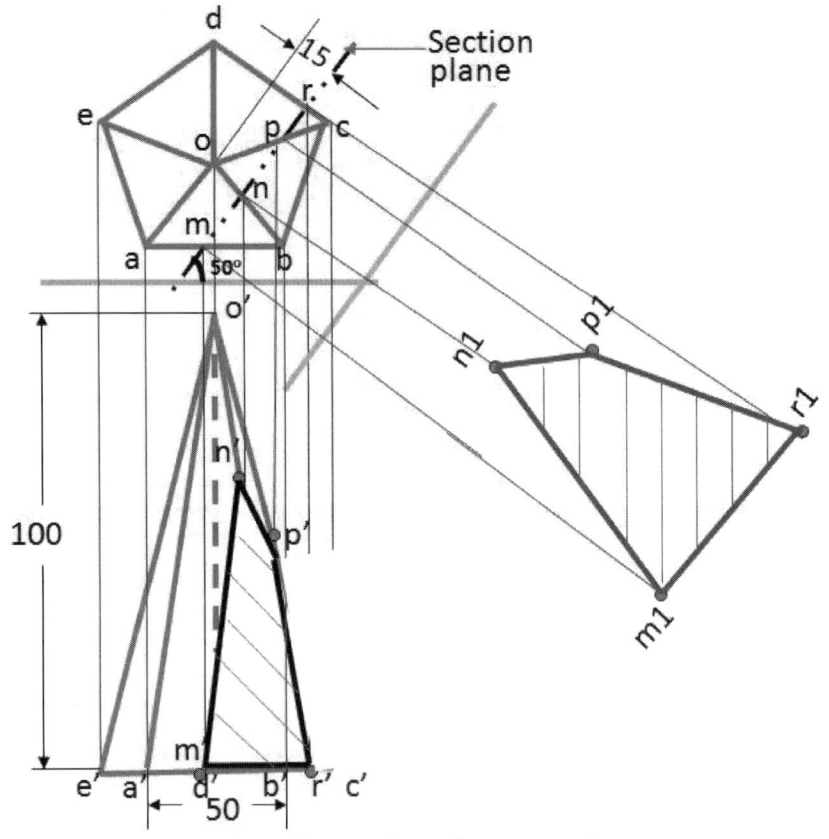

Figure 10-29: True surface of pentagonal section

The cutting plane cuts the axis of the pyramid 15 mm below the apex

It equally cuts the base at n and m

It cuts the edges at p and r

Join these points to O to form the section face

The true shape of the section (n₁m₁r₁p₁) is drawn as an auxiliary view to the top view with the reference line parallel to the section plane

Example

If the pyramid is also cut by another plane that is perpendicular to the frontal plane, inclined at 70° to the top plane and cuts the axis of the pyramid at 15mm from the apex. Draw the projections of the remaining part of the pyramid and the true shape of the cut section

Solution

Since the section plane is perpendicular to the frontal plane, the section line is drawn in the front view

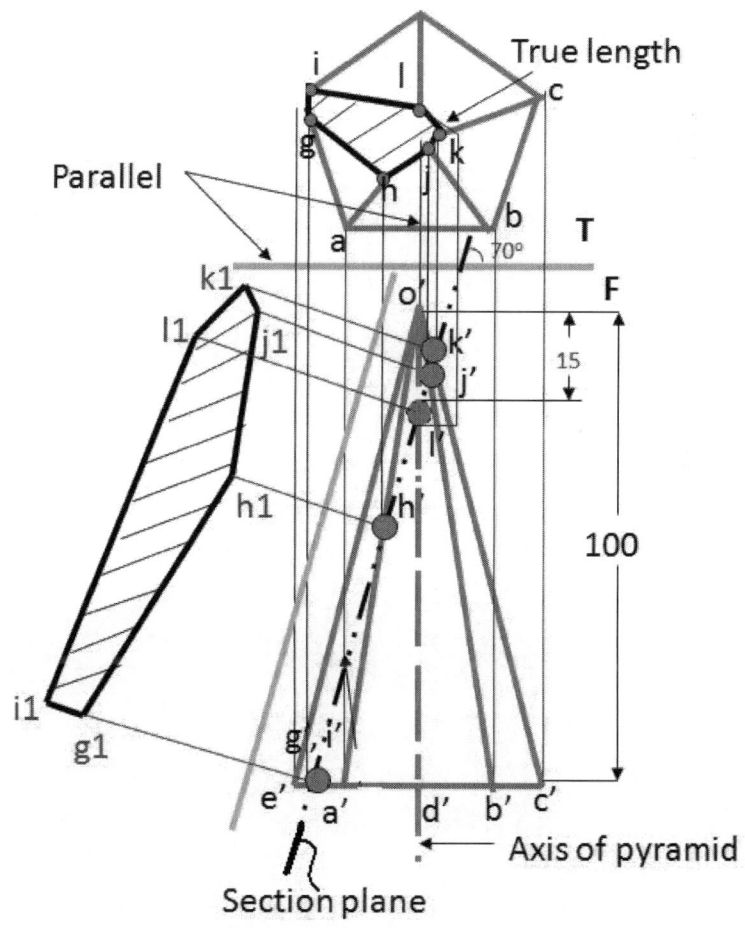

Figure 10-30: True surface of pentagonal section cut by frontal inclined plane

The cutting plane cuts the axis of the pyramid 15 mm below the apex

It cuts the base at g and i

It also cuts the edges at h, j, l and k

Project these points in the top view and join them

Eliminate the edge Oe and part of the edge Oa which are cut off

Project an auxiliary view of the true shape of the section by taking the reference line parallel to the section line

Locating a point on the sectional view

The plan and elevation of a heptagonal pyramid is shown in the figure below.

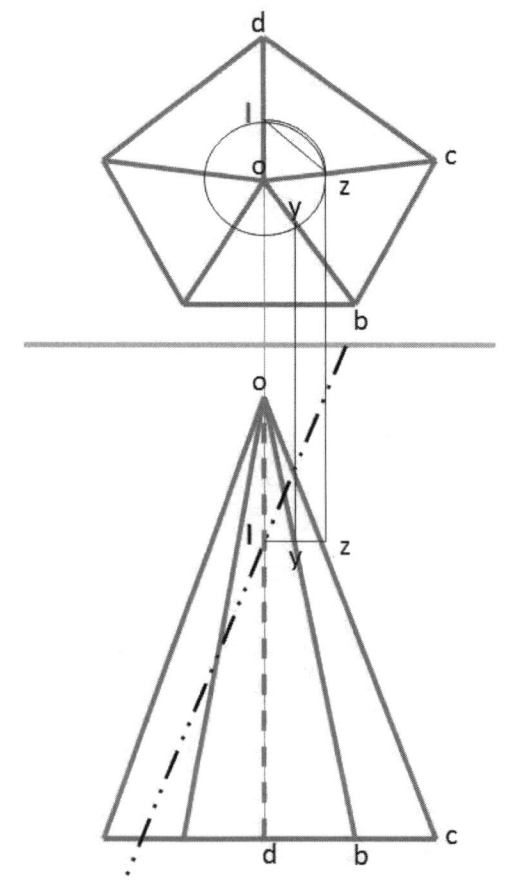

Figure 10-31: Locating point on a plane

To locate a point l on the plan view of the heptagonal pyramid, follow this process;

1. Draw an imaginary horizontal line from the axis (light blue) to the edge oc intersecting at z
2. Project the point z into the Top view (oz is TL here)
3. With o as center and oz as radius draw an arc cutting od at I
4. This can also be done by projecting onto ob at y and rotating.

Basically the imaginary line with length oz = oy is rotating inside the pyramid from one edge to another

This can also be obtained by drawing a line from z in the top view parallel to dc (as dc is TL here)

10.5 Surface development

Developable and non-developable surfaces

The process of laying out complete surface on one plane is called development of such surface. If the true size of all the faces of an object made of planes be found and joined in order along their common edges, the result will be the developed surface. Such surfaces which could be wrapped smoothly with a flexible material (such as paper) is said to be *developable*. Warped and double curved surfaces are *non-developable*, and when patterns are required for their construction they can be made only by some method of approximation, which is assisted by the pliability of the material will give the required form. Thus while a ball cannot be wrapped smoothly, a two-piece pattern developed approximately and cut from leather may be die-stamped, formed, or spun to a required shape.

Development of basic shapes

Many objects are constructed by cutting flat sheets to accurate dimensions, and by cutting, rolling or feeding then up into the famished shape after then may be secured at the joint. The joint maybe made by gluing, riveting, welding, brazing or bolting the edges together.

The following examples will illustrate the constructions methods involved in developing such surfaces. In the development of any object its projections must first be made, drawing only such views or parts of views as necessary to give the lengths of element and true size of cut surface.

10.5.1 Full surface development of geometrical solids

To develop a square pyramid

The development of a square pyramid is shown below in the form of four isosceles triangles. The base of each triangle is equal to the length of the base of the pyramid. The sides of the triangle have the same length as the edges of the pyramid as shown in Figure 9-25.

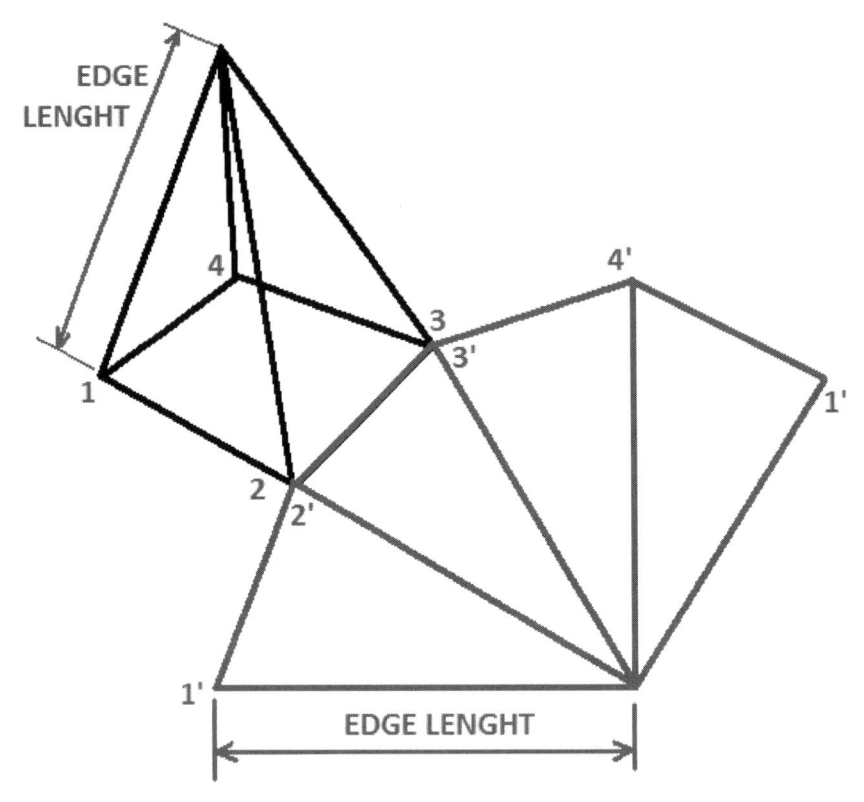

Figure 10-32: Surface development of pyramid

The development of the surface of the square pyramid consists of five surfaces of true shape laid out on a flat surface. As all development is true shapes it is necessary to find the true lengths of all surface edges.

1. With center A′ and radius A′1 draw the arc shown to a horizontal brought out from A′.
2. Project this point to the XY line and draw to A. this gives the true length of the edge of the pyramid.

Development of surface

1. With A as center and A1' as radius describe the arc.
2. On this arc step off the length of the base taken from the plan at 1', 2', 3', 4'.
3. Join these points to A to give the development of all four sides of the pyramid.

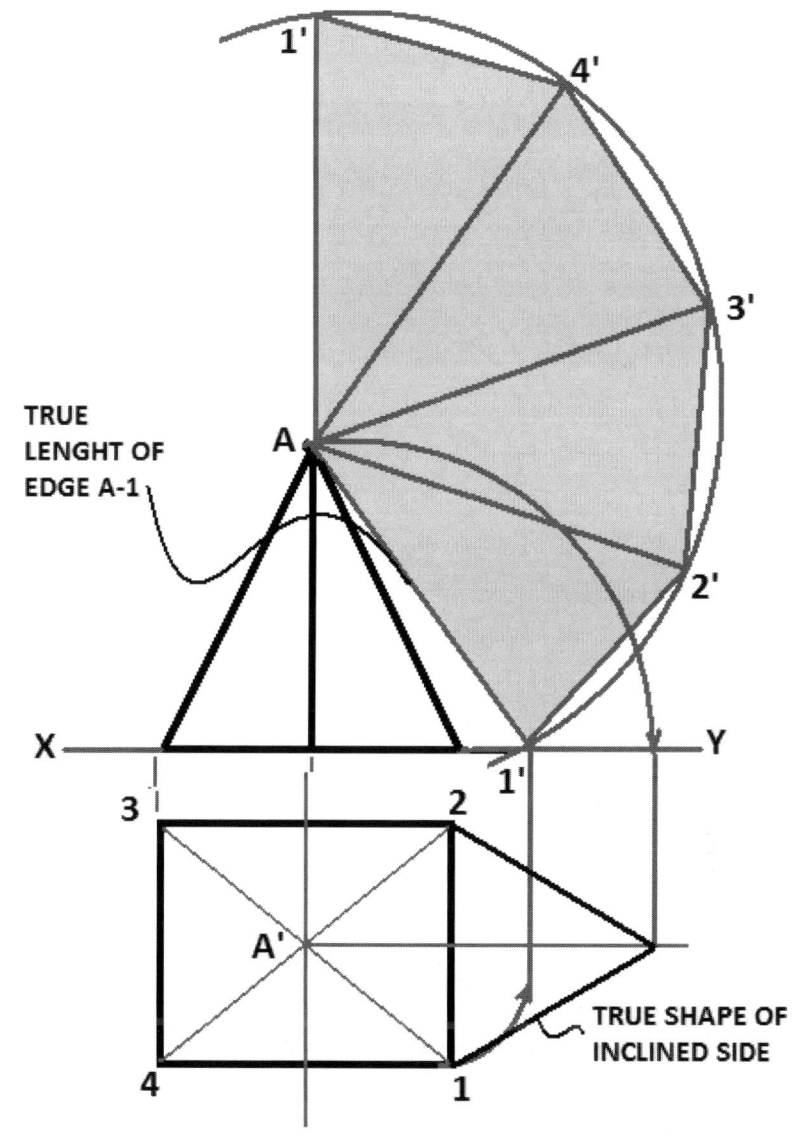

Figure 10-33: Surface development and true shape of pyramid

A second method showing the development of a single surface of the Figure 10-34 is also given below.

1. Draw the plan and elevation.
2. With C as center, and radius AC describe the arc to give A' on BC produced.
3. Project A' down to meet a horizontal from A' in A''.
4. Join A'' -3 -2 to give the development of the side.

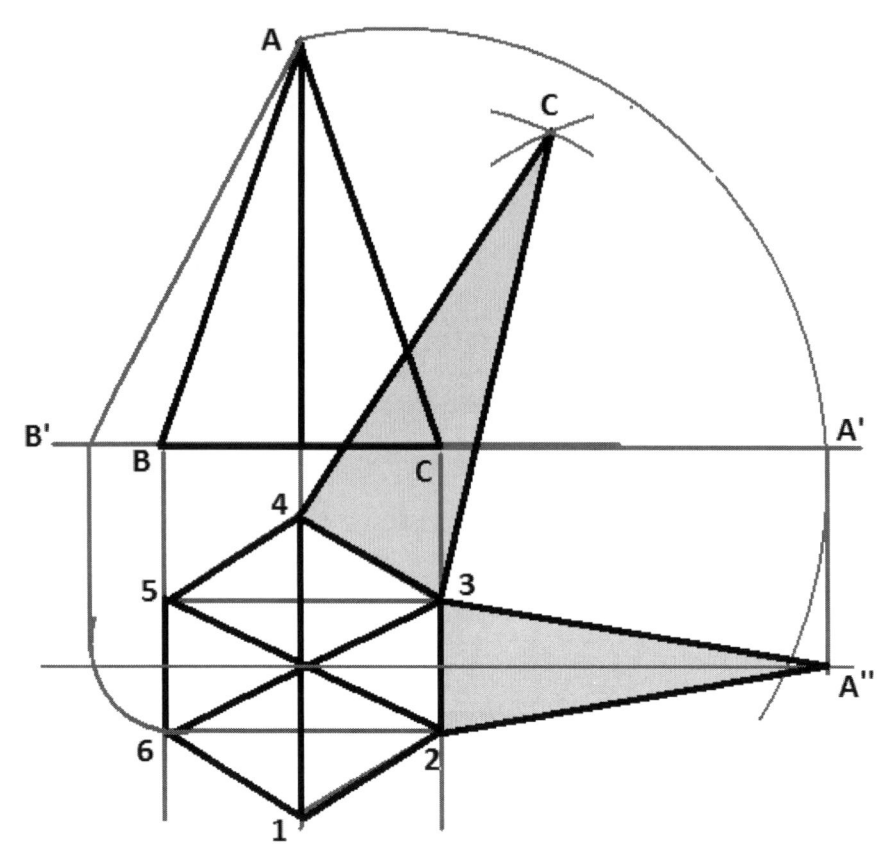

Figure 10-34: Surface development a hezagonal pyramid

It is important to note that a polygon on flat base has the true length as the slant edge projected and if the polygon is on one edge, the length of the slanting side on elevation is the true length.

10.5.2 Development of lower surfaces of geometrical solids

Frustums

Lower surfaces of cut pyramids (frustums) can be developed in the following ways considering the examples given below.

Example 1

1. With A as center and radius *AB* describe an arc and set off the length of the side (40mm) six times, obtaining the development of the six triangle faces.
2. Draw horizontal lines from points 2, 3 and 4 to give 2′, 3′, 4′.

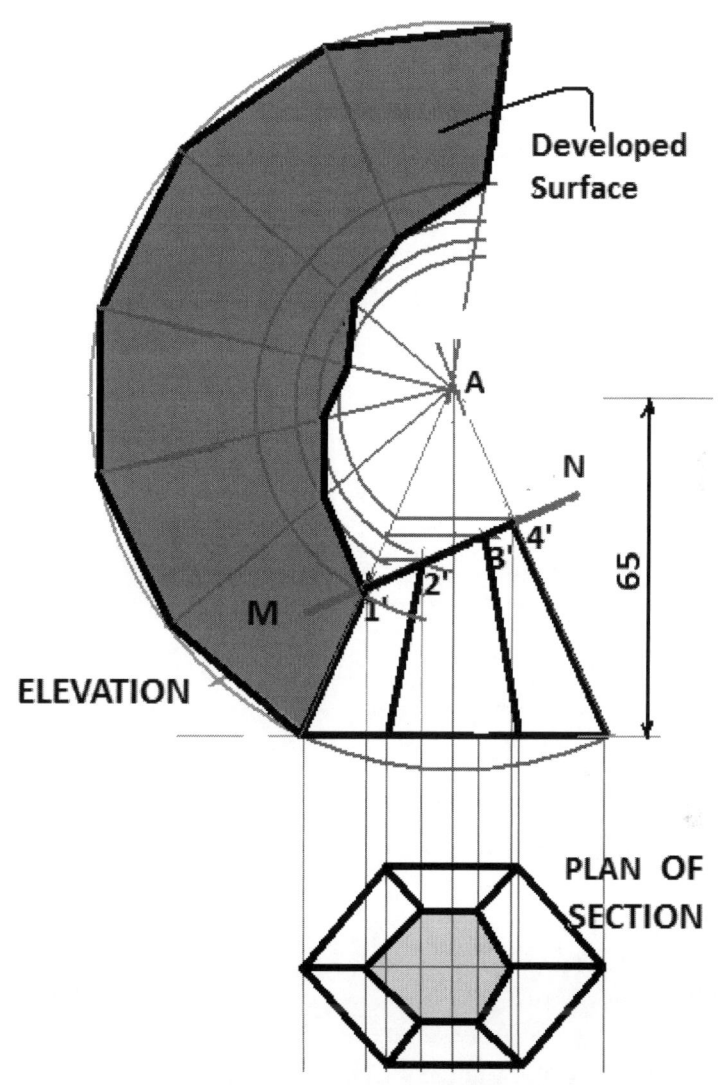

Figure 10-35: Surface development an irregular hezagonal frustrum

3. With A as center draw arcs from these points to cut the developed edges of the surfaces.
4. Join the points of instrument to give the development of the surfaces as shown.

Example 2

The development of a hexagonal pyramid made by an inclined plane is shown below. The true shape of the section is shown projected clear to the right on a new cutting plane M'N', onto which points brought out horizontally from the elevation are projected.

The dimensions *a* and *b* arc plotted on each side of a center line, the dimensions being taken from the plan view of the section.

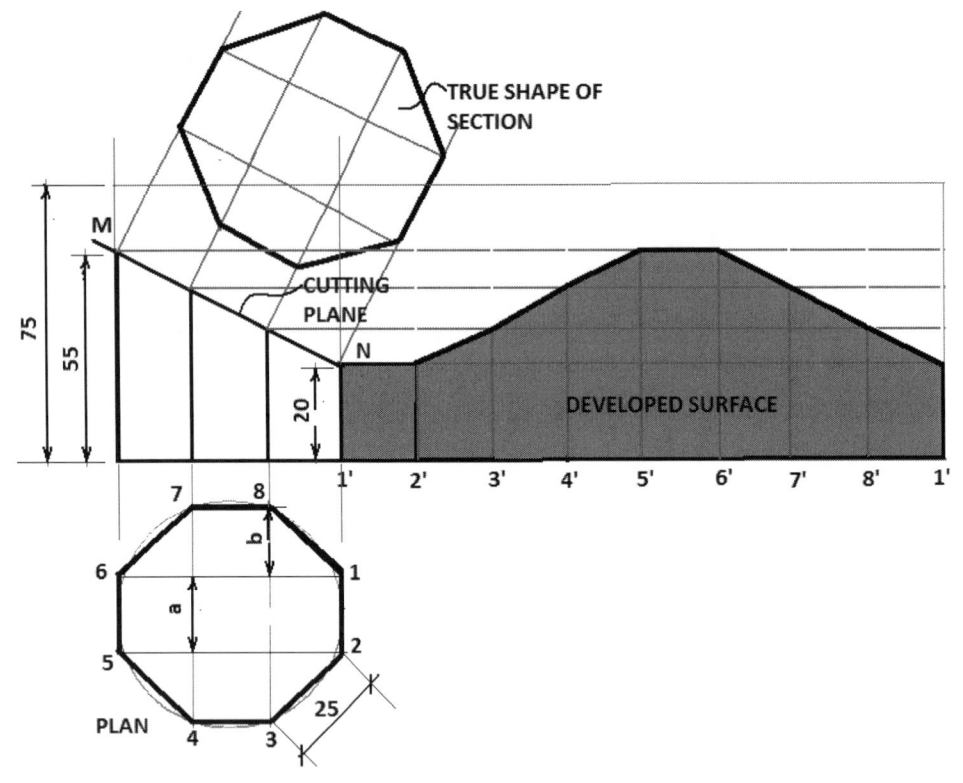

Figure 10-36: Surface development a regular hezagonal frustrum

Development of a right cylinder

In rolling the cylinder out on a tangent plane, the base, being perpendicular to the axis, will develop into a straight line. Divide the base into a number of equal parts, each representing elements. Project these elements up to the front view. Draw the stretch-out and measuring lines as before. Transfer the lengths of the elements in order either by projection or with dividers, and join the points by a smooth curve. Sketch the curve very lightly with freehand before fitting the curved ruler to it.

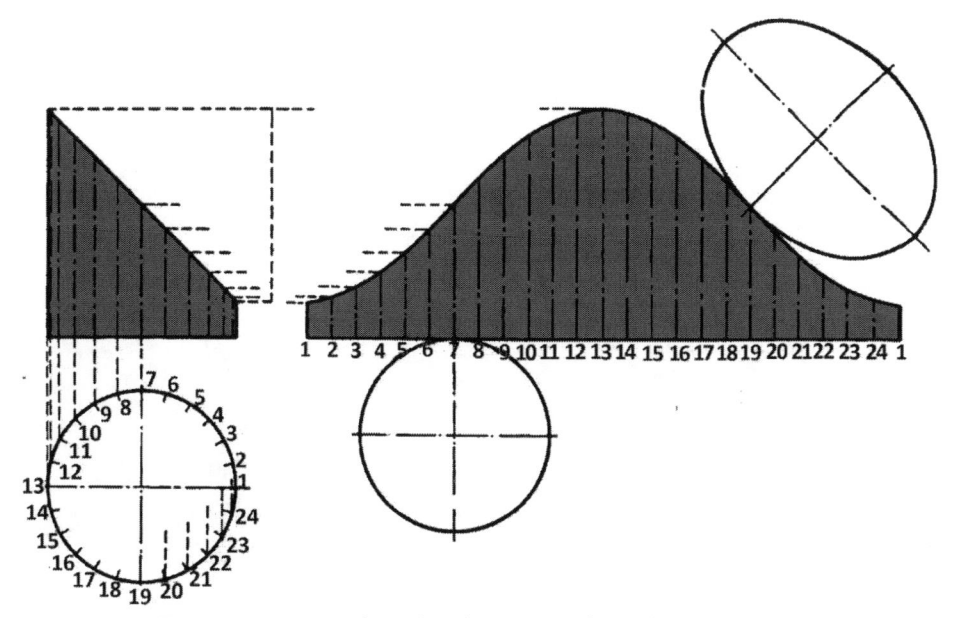

Figure 10-37: Surface development of a right cylinder

Development of frustum of cone

When a cone is cut by a cutting plane either parallel or inclined to its base, the portion remaining is called the frustum. The development of the base cut by a parallel cutting plane is shown.

The plan and elevation of the frustum of the cone the inclined plane MN and the development of the lower surface is shown below:

1. Draw the plan and elevation.
2. Divide the plan view into a convenient number of equal parts (12 parts in this case), project these points in the front view and draw the elements thought them.
3. With a radius equal to the slant height OA of the full cone, found from the contour elements which show the true length of all the elements, draw an arc, and lay off on it the divisions of the base, obtained from the plan view.
4. Connect these points with OA1 given the developed positions of the elements.
5. Find the true length of each element from vertex to cutting plane MN by revolving it to coincide with the contour element O^{v}A^v, and mark the distance on the developed position.
6. Draw the smooth curve through these points.

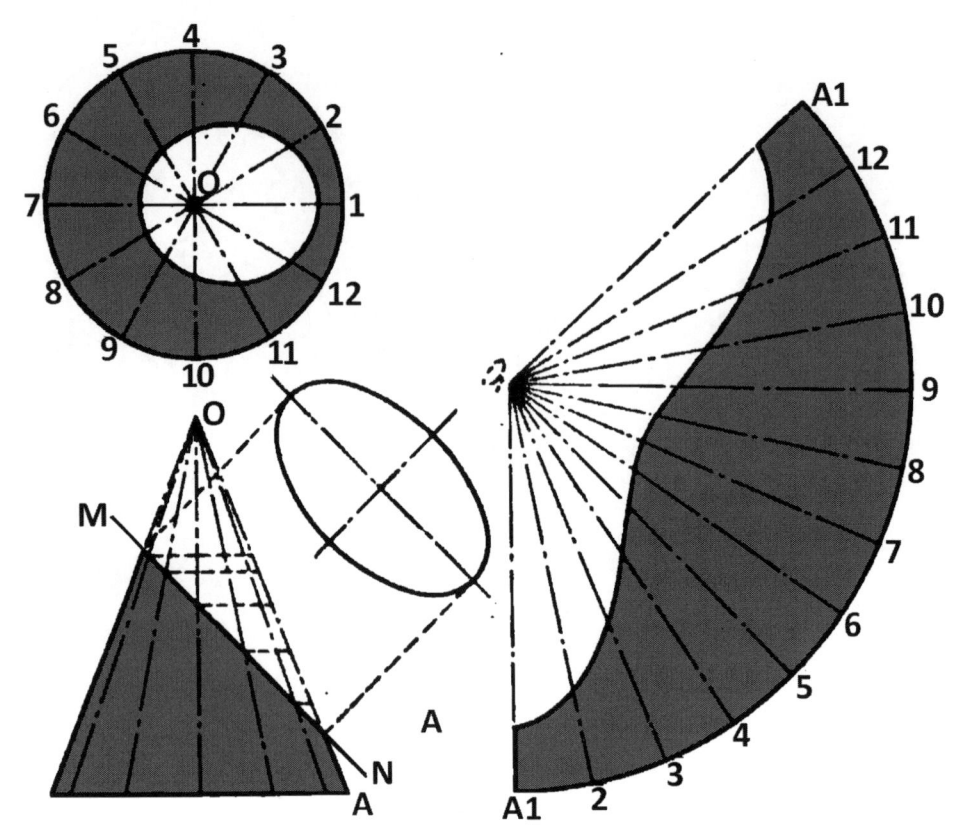

Figure 10-38: Development of right cone

The Figure 9-32 below showed the plan and elevation with the development of the frustum of the conical hood. The drawing procedure follows previous work on cones which reference should be made.

Triangulation

Triangulation is a process of dividing a given surface into small strips of triangles across the full length of the object. The commonest and best method for the development of non-developable surfaces is through the method of triangulation. This method is carried out by assuming the surfaces to be made up of narrow sections of developable surfaces i.e. the surface is assumes to be made up of large number of triangular strips, or plane triangles with very short bases. This method is used for all warped surfaces, oblique cones etc.

The principle consists of dividing the surface into small triangles, finding the true lengths of the sides of each triangle, and, constructing them at one time, and then joining these triangles on their common sides.

Example: Development of an oblique cone

An oblique cone differs from a right cone in that the elements are all of different lengths. The development of the right cone was practically made up of a number of equal triangles meeting at the vertex, whose sides were elements and bases the chords of short arcs of the base of the cone. In the oblique cone each triangle must be found separately.

Divide the base into a number of equal parts 1, 2, 3, (as the plan is symmetrical about the axis OC one-half only need be constructed). If the seam is to be on the short side the line OC will be the center line of the development and may be drawn directly at O_1C_1 as its true length is given at O^vC^v.

Find the true lengths of the elements O_1, O_2, O_3... by revolving them until parallel to V. The true length of any element is the hypotenuse of a right triangle whose altitude is the altitude of the cone and whose base is the length of the H projection. Thus to find the true length of O_1, lay off O^hC^h at D_n1_n

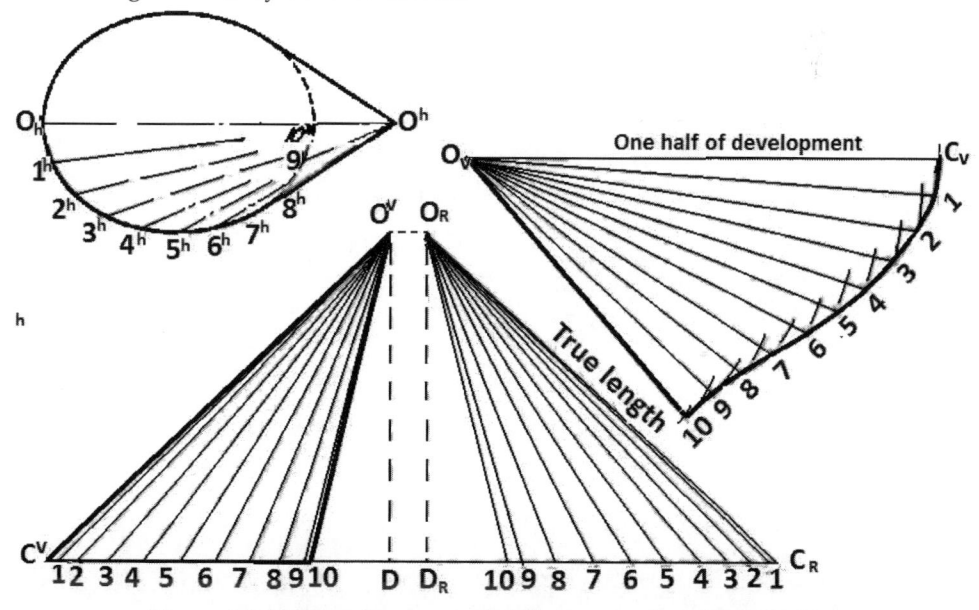

Figure 10-39: Development of oblique cone by triangulation

With O_1 as centre and radius D_h1_h, draw an arc on each side of O_hC_h. With C_1 as center and radius C^hI^h intersect these arcs at 1_1, then O_11_1 will be the developed position of the element O_1. With 1_1, as center and arc 1^h2^h intersect O_12_1 and continue the operation.

10.5.3 Interpenetration of surfaces

Intersection of surfaces

When two surfaces intersect, the line of intersection between them, which is a line common to both, may be thought of as a line in which all the elements of one surface pierces the other surface. Practically, every line on a drawing is a line of intersection; generally the intersection of two planes, such as a cylinder, cut by a plane, produces a circle. The term 'intersection of surfaces' refers to the more complicated line occurring when geometrical surfaces such as lines, planes, cylinders, cone, prisms, etc intersect each other.

Interpenetration of plane surfaces

There are two methods that can be used in constructing two surfaces that cut through each other;

a. A line piercing a surface and
b. A plane intersecting another plane

1. Construction of a line piercing a plane

This is possible in two methods

a. Edge view method and
b. Cutting plane method

Construction of a line piercing a plane using edge view method

Consider a line T_1T_2 penetrating a plane ABC. The views are shown in elevation plan and the side view in Figure 10-40 below. The elevation is drawn and the points projected to produce the plan and the side vies

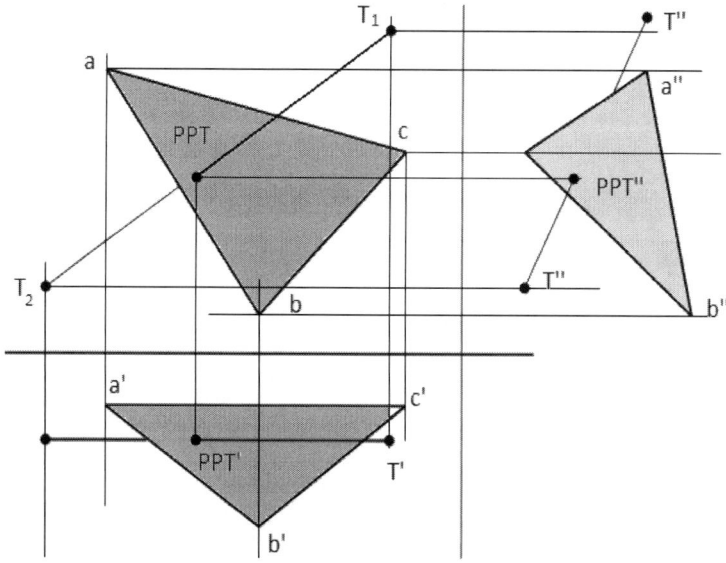

Figure 10-40: Edge view method

Construction of a line piercing a plane using cutting plane method

Consider a line penetrating a surface ABC shown in elevation and plane in Figure 10-41. Using the cutting plane method, the position of the penetrating line is shown in Figure 10-41 below

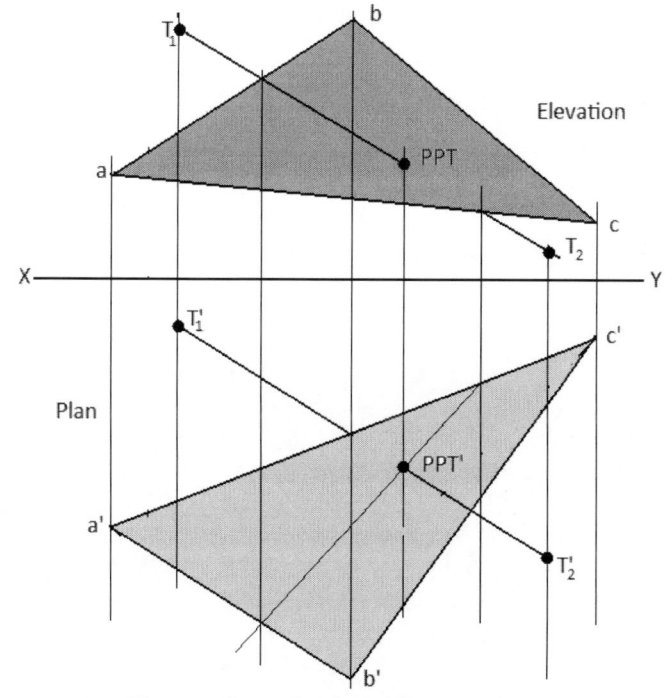

Figure 10-41: Cutting plane method

2. Construction of interpenetrating planes

This can also be achieved in two ways

c. Edge view method and
d. Cutting plane method

Construction of interpenetrating planes using edge view method

Consider two interpenetrating figures ABCD (a rectangle) and STU (a triangle) on a plane. The views (elevation and plan) are represented in Figure 10-42 below.

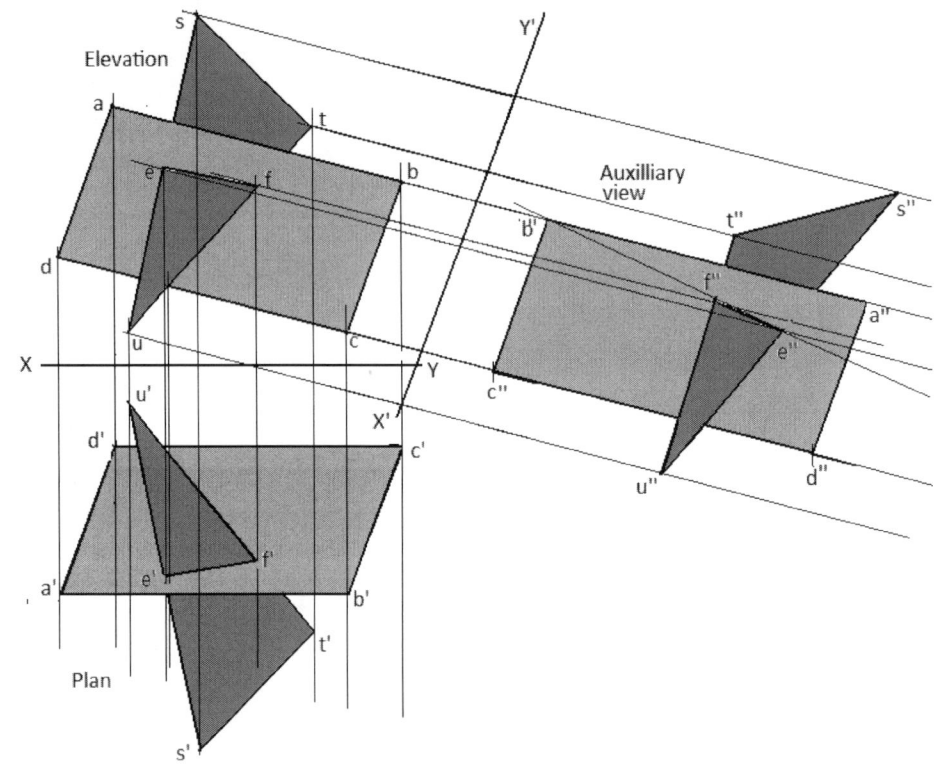

Figure 10-42: Edge view method

To develop the auxiliary view shown to the right of the figure, proceed as follows:

1. Draw the elevation view as shown and take projections of the plan view down.
2. Draw an auxiliary plane X'Y' parallel to one of the edge line of the rectangle.
3. Take projections from the elevation perpendicular to the produced X'Y' plane
4. Transfer true measurements from the elevation (vertical measurements) and the plan (horizontal measurements)
5. Complete the drawing and shade the planes

Construction of interpenetrating planes using cutting plane method

Consider two interpenetrating figures ABCD (a rectangle) and STU (a triangle) on a plane. The views (elevation and plan) are represented in Figure 10-43 below. To develop the view shown using the cutting plane method, an auxiliary view is not required, proceed as follows:

1. Draw the elevation view as shown and take projections of the plan view down.
2. Complete the drawing and shade the planes

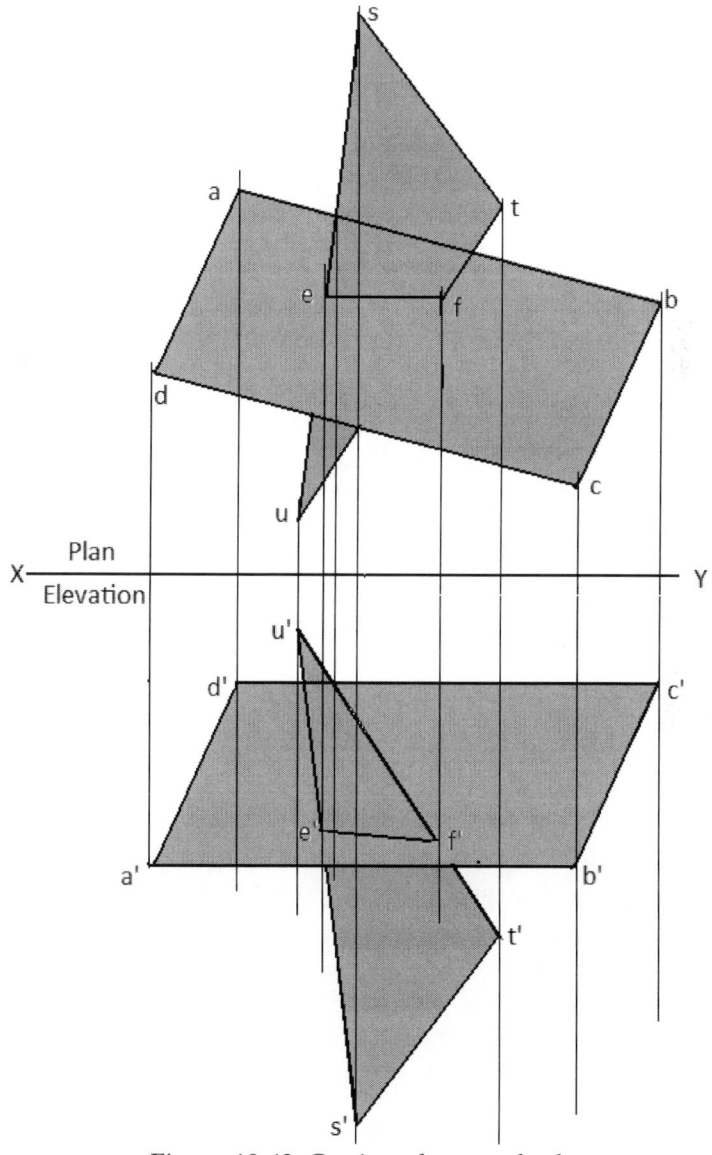

Figure 10-43: Cutting plane method

Intersection of solid figures (surfaces

a. Intersection of two cylinders

To find the intersection of two cyliders in position shown in 3-dimensions, the planes A, B, C, and D, parallel to V and shown in the same relative position on top and end views, cut elements from each cylinder, the intersection of which are points on the curve. The pictrial sketch shows a section of the planes while the development of the upper cylinder is evident from the figure.

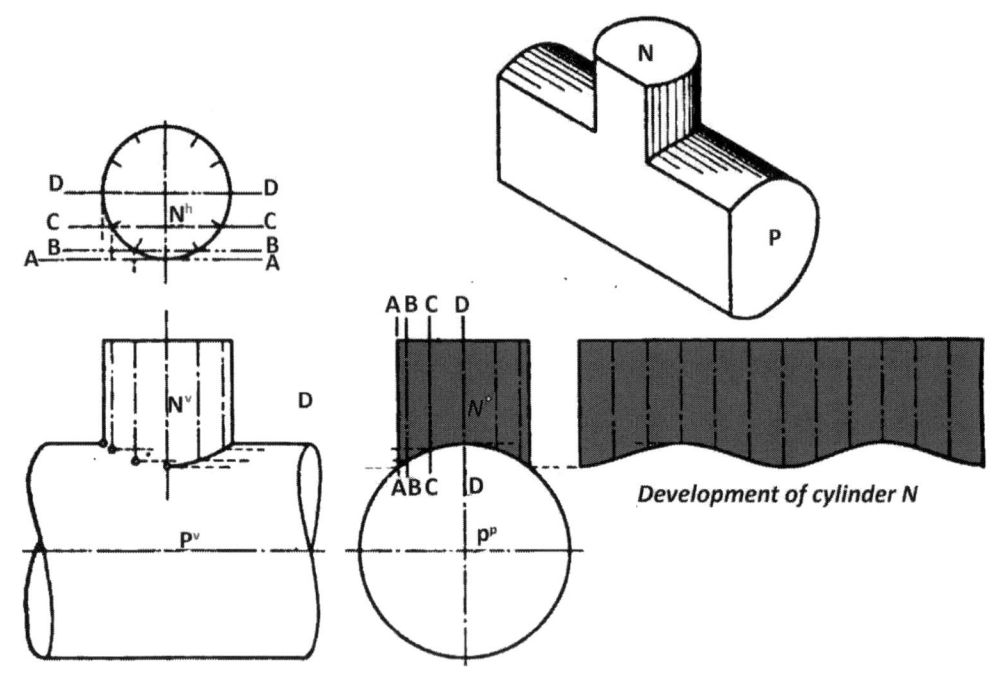

Figure 10-44: Intersection of two cylinders

b. Intersection of two prisms

To find the intersection of a triangular and square prism, the triangular prism is expected to pass through the square prism thereby creating two closed curves of intersection. A plane A-A parallel to the vertical plane through the front edge of the triangular prism cuts two elements from the square prism.

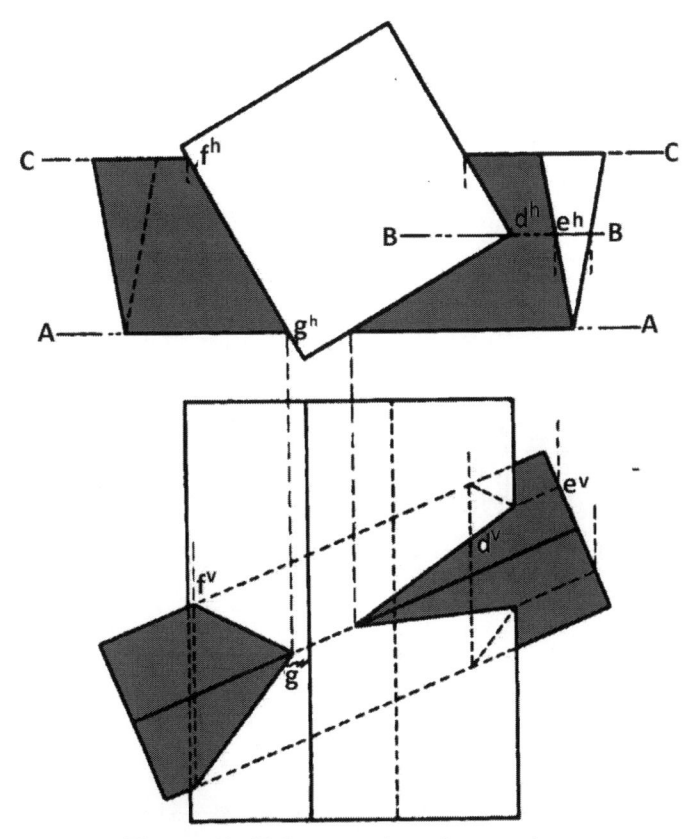

Figure 10-45: Intersection of two prisms

The front view shows where these elements cross the edge of the triangular prism, thus locating one point on each curve. The plane C-C contains the other two edges of the triangular prism and will give more points on each curve. As on the left side only one face of the square prism is penetrated, the curve would be a triangle, two sides of which are visible and one invisible. On the right side two faces are penetrated. The plane B-B is thus passed through the corner, the two elements cut from the triangular prism projected to the front view, where they intersect the corner as shown.

Exercise

1. Consider the diagrams below and select the correct sectional view

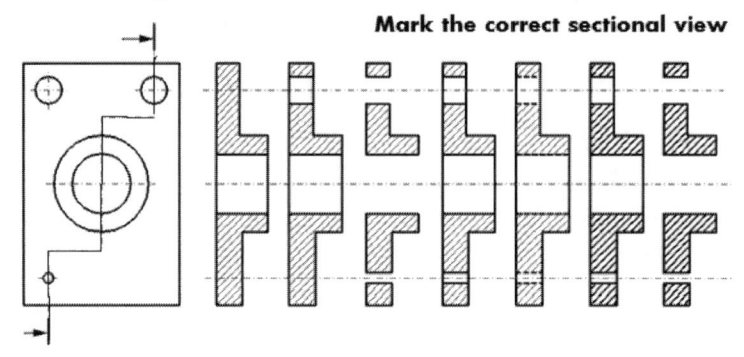

   Mark the correct sectional view

2. Draw the auxiliary plan of these geometrical solids with the auxiliary planes inclined at given angles to XY plane

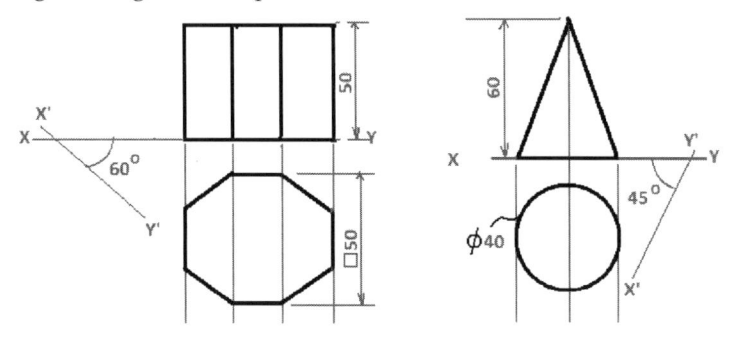

3. Draw the auxiliary elevation and plan of the octagonal pyramid shown below.

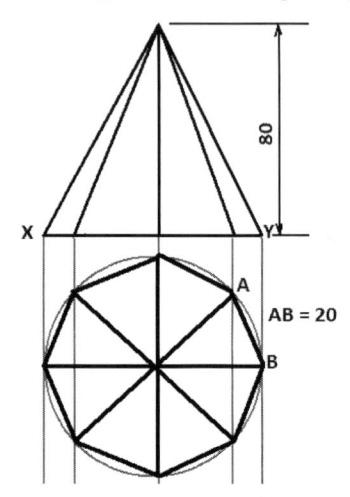

   AB = 20

4. Find the line of intersections in the figures above

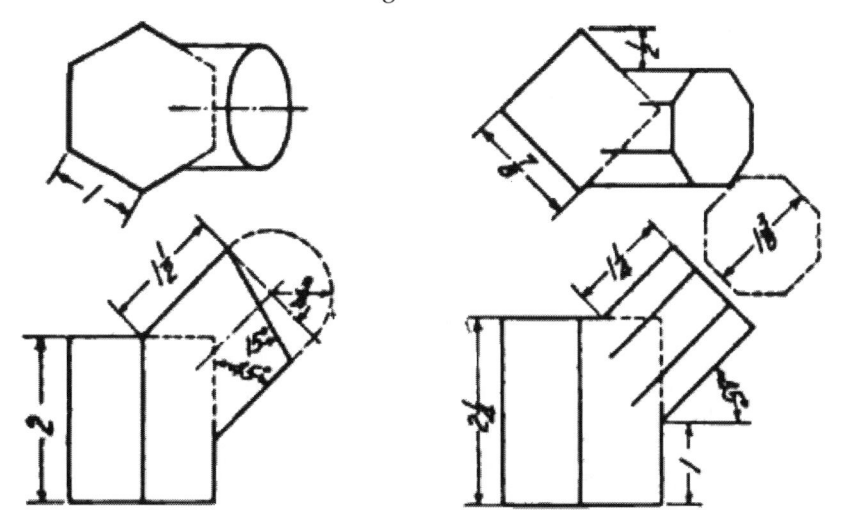

References

Bellis H.F., Schmidt W.A., *Architectural Drafting,* New York, McGraw - Hill Book Co., 1971.

David Anderson, 2006. Technical drawing. Class handout on MEC 1000 Spring 2006. Technical drawing-class handout.pdf. Date accessed 26/01/2012

Gülsev Uyar Aldaş 2008. JFM210 technical drawing and computer application lecture notes (first part). Technical drawing.pdf. Date accessed 3/6/2008

http://www.mkn.itu.edu.tr/~mkimrak/MAK112E_dersnotu.htm

Margo Gonterman, 2012. Assignment 6: Parabola Construction. Class note, Department of Maths and Sci. Educ. The University of Georgia, US

McBean G., Kaggwa N. and Bugembe J., *Illustrations for Development,* Nairobi, Afrolit Society, 1980.

MIT Open Course Ware: http://www.ocw.mit.edu/NR/rdonlyres/Mechanical-Engineering/2-007Spring-2005/929103E2-EBAD 40DE88BFE2258E0FEC49/0/drawings.pdf - 2006-11-09

Parker M. A. & Pickup F., 1976. Engineering Drawing with worked examples 1. Hutchinson & co (Publishers) Ltd, UK

Styles K., *Working Drawings Handbook,* London, Architectural Press, 1982.

Taylor R., *Model Building for Architects and Engineers,* New York, McGraw-Hill Book Co., 1971.

Technical Drawing Instruments & Their Uses. http://www.ehow.com/list_6762592_technical-drawing-instruments-uses.html#ixzz2EjOY97D5

Winden V., Keijzer M., Pforte W., Hohnerlein F., *Rural Building - Drawing Book,* Maastricht, Netherland, SticktingKongregatie F.I.C.

www.tech.plymouth.ac.uk/dmme/dsgn131/DSGN131_Course_Notes.pdf

Notes

Titles in author's list

The book gave an overview of agricultural engineering fundamentals, which is does not adequately represent some aspects of field practice in engineering training in our University, Polytechnic and Colleges curricular. This Volume 1 of the title series 'Agricultural Engineering principles & practice covers wider scope of agricultural engineering practice. Three major aspects of agricultural engineering were explored: Agricultural engineering development, Agricultural land preparation and Crop planting and establishment.

ISBN-13: 978-147-931-614-4

URL:https://www.createspace.com/3996235

Available on-line

This Volume explores engineering involvement in soil and water conservation, agricultural material properties, processing and handling as well as farm structure requirement, farmstead layout, storage structures and construction, animal housing requirements among others. The book undoubtedly provides essential engineering fundamentals required by students for effective teaching and practical training in skill acquisition. The book is therefore recommended for all students of agricultural and engineering technology students in training at different levels in the university, polytechnic, colleges and vocational schools.

ISBN-13: 978-145-633-568-7

URL: https://www.createspace.com/3498612

Available on-line

Fundamental Principles of AGRICULTURAL ENGINEERING PRACTICE

SEGUN R. BELLO

This book provide an overview of the advances which have been made and are currently in progress to provide a strong base for a review of agricultural engineering curriculum in order to catch up with the global trend in agricultural engineering revolution especially in Nigeria. For the ever increasing population, the drudgeries involved in food production, incurable losses in harvest and post harvest operations as well as the ever increasing and increased expectation of high quality food products meeting consumers' need and satisfying food safety standards had called for the growth of accurate, fast and objective quality determination indices of agricultural and cost effective techniques employed in food production.

ISBN-13: 978-080-015-8

Available in bookstores

The dynamic nature of agricultural operations and the complexity of agricultural machinery are indices of scientific research diversity as evident in the wide spread requirements in agricultural operation sustainable production. Engr. Segun presents extensive works on agricultural mechanization and machinery utilization in agricultural production documented in this eleven chapter book to acquaint students and researchers with the principles of agricultural machinery and provide them with requisite knowledge and skills on various agricultural machinery requirements for effective agricultural mechanization.

ISBN-13: 978-145-632-876-4.

URL: https://www.createspace.com/3497673

Available on-line

AGRICULTURAL MACHINERY & MECHANIZATION

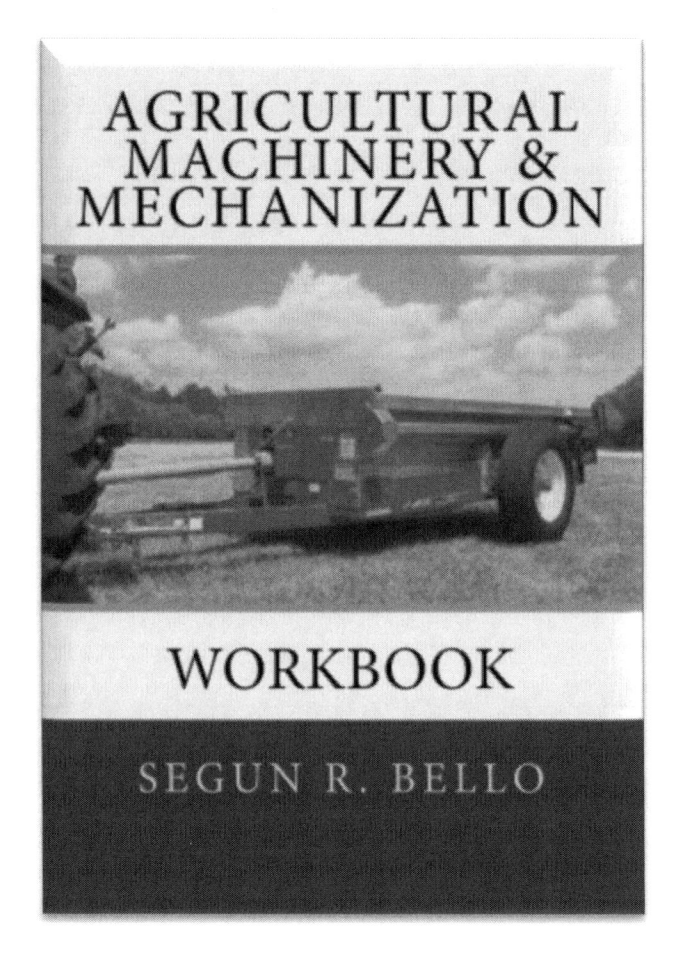

WORKBOOK

SEGUN R. BELLO

The author designs this workbook to help students have an understanding of the practical content of the agricultural machinery as a course and to guide them in carrying out determination of mechanization indicators, machine performance indices and also field experimentation, monitoring and reporting. This is in effort to improve the quality of practical presentation and documentation to meet the requirements of NBTE, NUC and other examination bodies. The workbook directly improves students' opportunity to learn new concepts of log entry and field measurement and computation by direct participation, acceptance of new methodology from instructors, and breeding of future technicians.

ISBN-13: 978- 1484927038

- 1484927036

URL:https://www.createspace.com/4277084

Available on-line

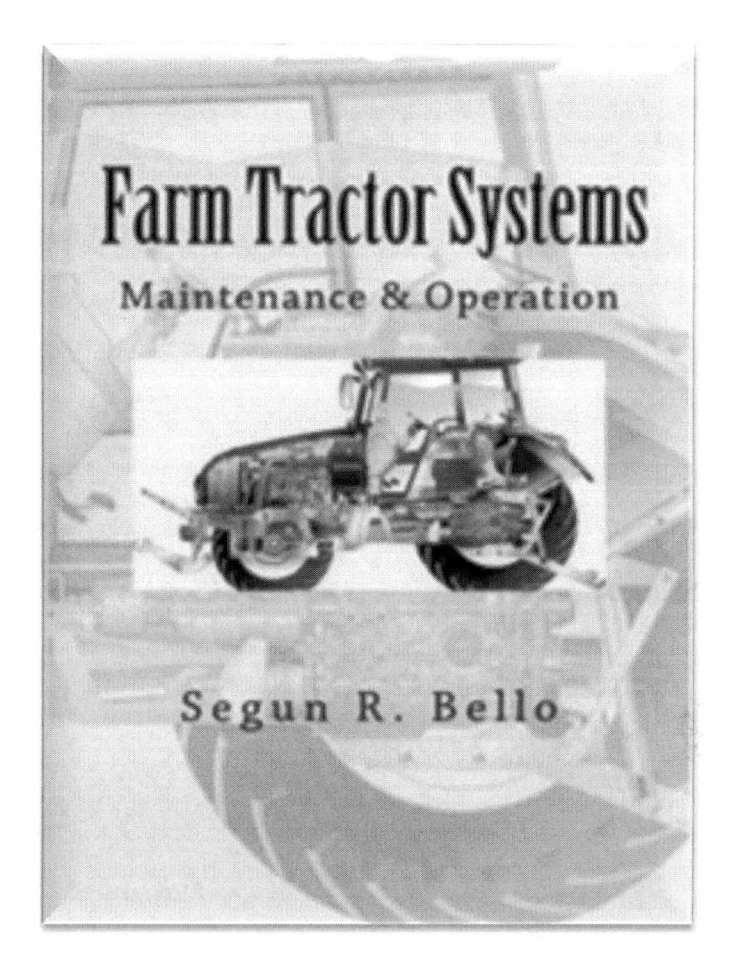

Farm Tractor Systems
Maintenance & Operation

Segun R. Bello

A link between machine functionality, operations, performance and decision making in the management of power sources and field operations were presented in this book. Depreciation and functional deviation of a machine from its original state at manufacture could put the life of a machine in danger of breakdown or obsolescence, which is counted a loss to any such organization or the entrepreneur. To avoid such losses, an understanding of machine systems functionality and a well organized maintenance programme designed to maintain, prevent or restore machine to near original state is required.

ISBN-13: 978-148-102-292-7

URL:https://www.createspace.com/3996235

Available on-line

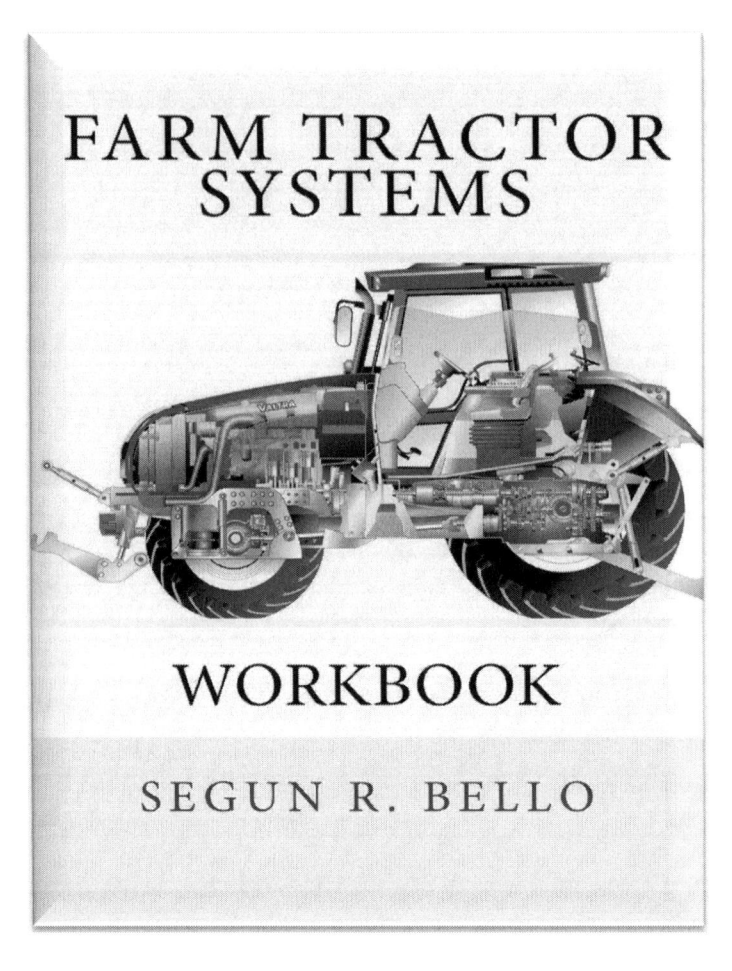

FARM TRACTOR SYSTEMS

WORKBOOK

SEGUN R. BELLO

The author designs this workbook to help students have an understanding of the practical contents of the farm tractor and to guide them in carrying out system maintenance, repairs, overhauling and engine tune-up as well as reporting field experimentation and monitoring to improve the quality of practical presentation and documentation in order to add value to quality. The practical exercises improve students' opportunity to learn new concepts by direct participation, acceptance of new material from instructors, and breeding of future technicians.

ISBN-13: 978-148-491-835-7

URL:https://www.createspace.com/4272459

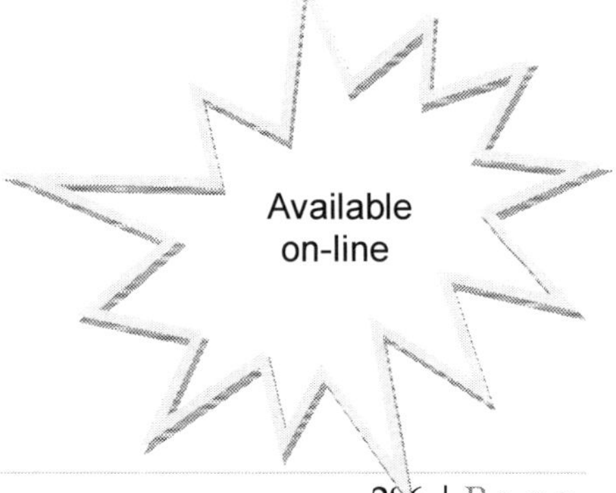

Available on-line

This book is all you need in emergency breakdown and where there is no mechanic. It offers a guide to decision making in machinery procurement, farm power selection, engine troubleshooting, tractor driving and operations as well as tractor and machinery maintenance and repairs. In this way, the enormous costs and valuable time spent on waiting desperately at breakdown points, tracing of faults, annoying breakdowns, unnecessary down time and costly repairs can be adequately reduced.

ISBN-13: 978-332-254-4-3

Available in bookstores

This Practical workbook is an expression of the student's desire to have a simplified extraction from the practical content of the topics discussed in my previous work; **Guide to Agricultural Machinery Maintenance and Operations**. There is urgent need for students to learn the art of presentation of technical report through active participation and reporting. This workbook present a simple approach to achieving such objective than it had been in the past. With the contents of this workbook, it is easier to follow laid down procedures to carry out practicals, and report them appropriately. Conducting, reporting and documentation of students' practical activities therefore become easier and more presentable.

ISBN-13: 978-376-67-1-6

Available in bookstores

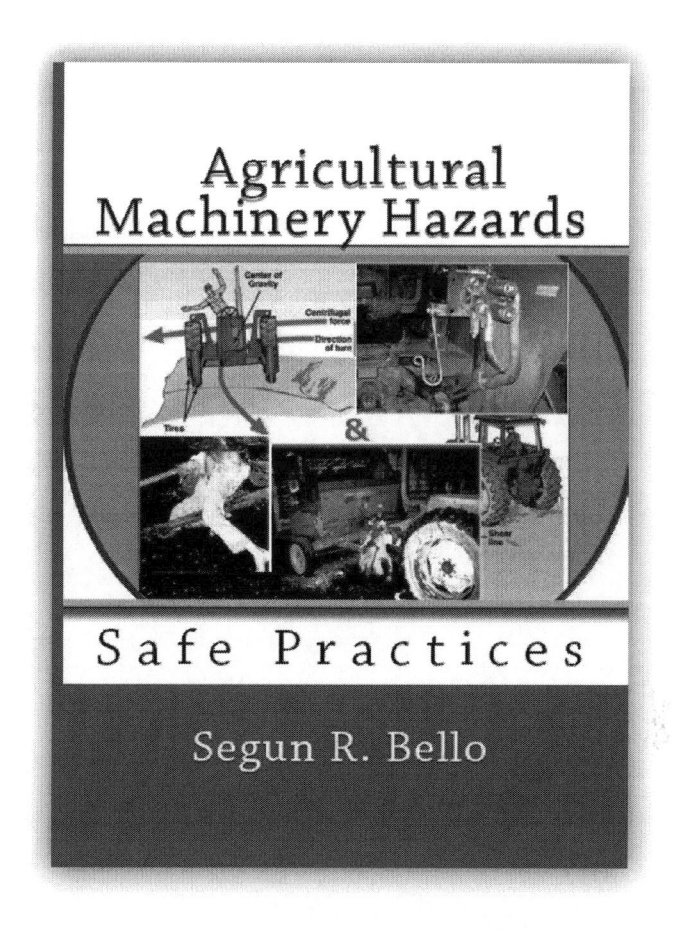

Agricultural Machinery Hazards & Safe Practices

Segun R. Bello

As long as agriculture underpins the survival of humanity, safety remains a relevant issue to life security in and around the farm community for system sustainability. An understanding of the issues and values of hazard and safety in machinery operations as presented in this book with *full coloured graphic prints* will aid in decision-making reinforced by principles and practice as well as facilitate effective utilization of signal communication techniques and the attainment of relevant knowledge in accident prevention in primary production processes.

ISBN-13: 978-146-790-718-7

URL:https://www.createspace.com/3498621

Available on-line in full colour

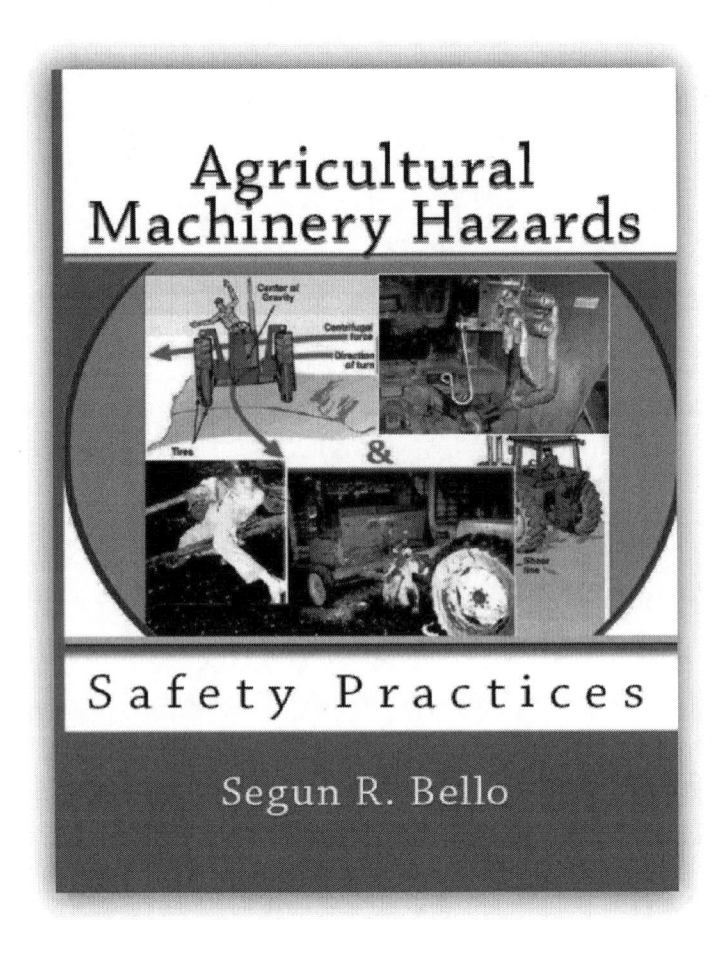

Agricultural Machinery Hazards & Safety Practices

Segun R. Bello

The diversity and complexity of agricultural and related machinery have become an index for increased rate of accident and injury occurrence experienced during operations and maintenance. Therefore, the study of machinery hazards, hazard sources and points in machinery and subsequent safe practices will help to eliminate, eradicate or control such hazards and provide workers with the opportunity to operate machinery more safely and develop skills in improved material and machine handling, as well as facilitate effective utilization of signal communication techniques and the attainment of relevant knowledge in accident prevention in primary production processes.

ISBN-13: 978-147-753-664-3

URL: https://www.createspace.com/3728177

Available on-line in black & white

WORKPLACE HAZARDS
Risks & Control

Segun R. Bello

In as much as we live within hazardous environments, it is our responsibility to make the environment favourable. It is our responsibility to provide guide to workers' safety, change attitude and offer safety training programmes to ensure safe work environment. Remember, it is important to make rules about safety; however, it is more important to ensure safety by locking dangers away. This book x-rays the various workplaces and associated hazards as well as provides an insight to some measures of safety within workplace.

ISBN-13: 978-147-528-554-3.

https://www.createspace.com/3865653

Available on-line

This book is designed to help students acquire requisite knowledge and skills in basic workshop technologies & practices, workshop management, organization and handling of tools and machines in preparations to meet the demands of the manufacturing and processing sector of our economy. The author believed that reading through this book, users will be able to appreciate the work environment and the influences it has on the workers' safety and as well have gained enough experience that will guide you in safe tool handling and machine operation which guarantees effective job delivery without incidences of hazards, injury or accident.

ISBN-13: 978-147-928-308-8

URL: https://www.createspace.com/3982311

Available on-line

This book was packaged to help students acquire requisite knowledge and practical skills in engineering/technical drawing practices. The contents were designed to prepare students for technical, diploma and degree examinations in engineering, engineering technology and technical vocations in other professions in the monotechnics, polytechnics and universities. Emphasis is placed on media drafting, lettering, and alphabet of lines, geometric construction, sketching, and multiview drawings.

ISBN-13: 978-148-125-012-2

URL:https://www.createspace.com/3996235

Available on-line

Fruits and Vegetable Technologies
Segun R. Bello

This book is written to provide the students with a good understanding in fruits and vegetables handling, processing, and technological advances in preservation of fruits and vegetable from harvest t.ill it gets to the consumer table or ended at the store shelf as finished products. Fruits and vegetables surfers the highest degree of deterioration at all levels of technological involvement right from maturity till shelving. This book is therefore packaged to advance knowledge and increase understanding of the nature of the fruits and vegetables in order to match up the principles and techniques of crops handling, processing and storage in order to minimize post harvest losses.

ISBN-13: 978- 149-047-910-1

-10: 149-047-910-4

URL:https://www.createspace.com/

Available on-line

HORTICULTURAL MACHINERY

Equipment & Safety

Segun R. Bello

This book is packaged to provide the students with background knowledge of various horticultural operations, tool and equipment use. Written in simplified English with detailed graphic illustrations and pictures, the book is the perfect tool required in every home to in selecting tools and machines for horticultural and gardening operations.

ISBN-13: 978- 148-497-487-2
    -10: 148-497-487-5
URL:https://www.createspace.com/4284225

Available on-line

HORTICULTURAL MACHINERY WORKBOOK

SEGUN R. BELLO

The author designs this workbook to help students have an understanding of the practical content of the horticultural machinery course and to guide them in reporting field experimentation and monitoring. This is in effort to improve the quality of practical presentation and documentation in order to add value to quality of practical as well as improve students' opportunity to learn new concepts by direct participation, acceptance of new instructional materials, and breeding of future technicians.

ISBN-13: 978-148-492-821-9
       -148-492-821-0
URL:https://www.createspace.com/4277259

Available in bookstores

Guide To Agricultural Machinery Maintenance and Operations

For Engineering Students, Agricultural Technologists & Tractor Operators

Segun R. Bello

ISBN: 978-2986-90-9
FERP - FASMEN

This manual is prepared to provide an essential guide to students' practical in agricultural engineering and agricultural technology programmes and also at appropriate levels in other tertiary institutions in the country. In preparing the manual, the requirements and minimum standards specified by the various academic regulatory bodies in Nigeria such as: National Board for Technical Education (NBTE), Nigeria Universities Commission (NUC) Nigeria Society of Engineers (NSE), National Commission for Colleges of Education (NCCE), Council for the Regulation of Engineering in Nigeria (COREN) etc, were taken into consideration.

ISBN-13: 978-298-6-90-9

Available in bookstores

Chapter contribution in Books

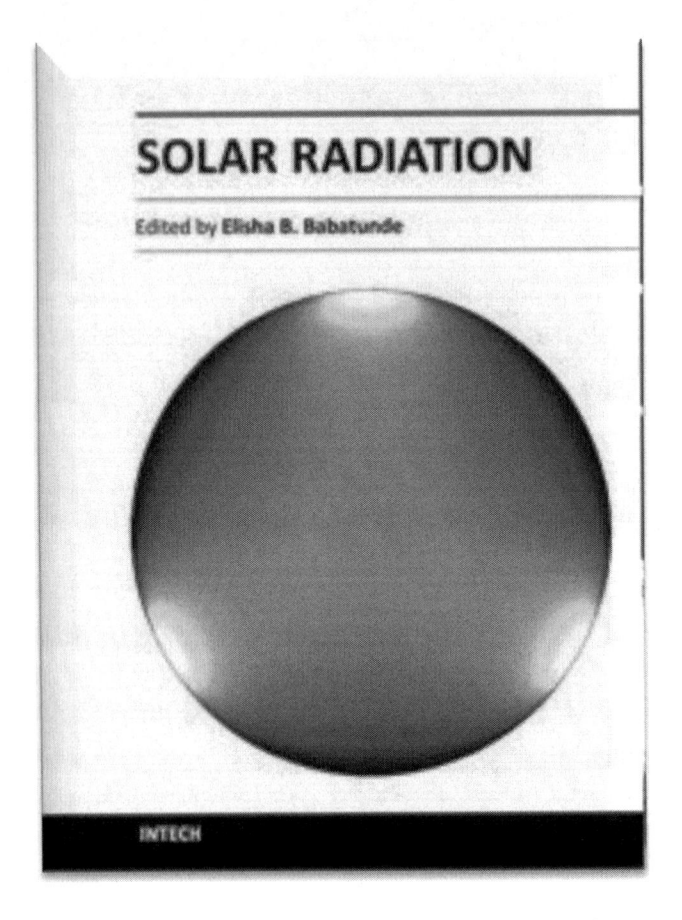

The book presents fundamental and well researched contributions on possible, feasible and future applications of solar radiation as an energy source by world class scientists including the author. As old as its source, the sun, little did the world knew of its potential as an enormous energy provider. It has now attracted the attention of scientists, engineers and even the public and attracted the attention of the academic curricula of science and engineering courses in higher institutions. It is studied as an environmental science and as an energy course, particularly in the aspect of alternative or renewable energy source both in science and engineering departments of universities.

ISBN: 978-953-51-0384-4.
http://www.intechopen.com/books/solar-radiation

Available on-line

Edited books

Sustainable Agriculture
Challenges & Prospects

Edited by Segun R. Bello (Engr), Balogun R. Babatunde

Sustainability of agricultural production system is becoming a major concern to agricultural research and policy makers in both developed and developing countries as it represents the last step in a long evolution of the protection of natural resources and the maintenance of environmental quality. This 6-part book furnish scientists and students with fundamental views on scientific developments, research outcome on sustainable solutions and also offers guidance on dissemination of sustainable agricultural techniques and feasible applications to Nigeria situation as a way of wriggling out of the ever expensive, environmentally degrading conventional machine and inorganic agricultural production practices.

ISBN-13: 978-148-010-344-3

URL: https://www.createspace.com/4025911

Available on-line

For more information, visit:

1. http://www.amazon.com/Segun-R.-Bello/e/B008AL6RI0
2. http://www.amazon.com/s?ie=UTF8&field-author=Engr%20Segun%20R.%20Bello&page=1&rh=n%3A283155%2Cp_27%3AEngr%20Segun%20R.%20Bello
3. http://www.amazon.com/Segun-R.-Bello/e/B008AL6RI0
   http://www.amazon.com/s?ie=UTF8&field-author/
4. http://lejpt.academicdirect.org/
5. http://www.cigr-ejournal.tamu.edu/
6. http://www.intechopen.com/books/solar-radiation
7. http://www.medwelljournals
8. http://www.sciacademypublisher.com/journals/index.php/SATRESET